T0261604

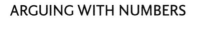

ARGUING WITH NUMBERS

RSA·STR

THE RSA SERIES IN TRANSDISCIPLINARY RHETORIC

The RSA Series in Transdisciplinary Rhetoric is a collaboration with the Rhetoric Society of America to publish innovative and rigorously argued scholarship on the tremendous disciplinary breadth of rhetoric. Books in the series take a variety of approaches, including theoretical, historical, interpretive, critical, or ethnographic, and examine rhetorical action in a way that appeals, first, to scholars in communication studies and English or writing, and, second, to at least one other discipline or subject area.

Edited by James Wynn and G. Mitchell Reyes

ARGUING WITH NUMBERS

The Intersections of Rhetoric and Mathematics

THE PENNSYLVANIA STATE UNIVERSITY PRESS

UNIVERSITY PARK, PENNSYLVANIA

Library of Congress Cataloging-in-Publication Data

Names: Wynn, James, 1972– editor. | Reyes, G. Mitchell, editor.
Title: Arguing with numbers : the intersections of rhetoric
    and mathematics / edited by James Wynn and
    G. Mitchell Reyes.
Other titles: RSA series in transdisciplinary rhetoric.
Description: University Park, Pennsylvania : The Pennsylvania
    State University Press, [2021] | Series: The RSA series in
    transdisciplinary rhetoric | Includes bibliographical
    references and index.
Summary: "A collection of essays which deploy rhetorical
    lenses to explore how mathematics influences the values
    and beliefs with which we assess the world and make
    decisions, as well as how our values and beliefs influence
    the kinds of mathematical instruments we construct
    and accept"—Provided by publisher.
Identifiers: LCCN 2021001829 | ISBN 9780271088815 (cloth)
Subjects: LCSH: Mathematics—Social aspects. | Rhetoric—
    Social aspects. | LCGFT: Essays.
Classification: LCC QA10.7 .A64 2021 | DDC 303.48/3—dc23
LC record available at https://lccn.loc.gov/2021001829

The Pennsylvania State University Press is a member of the
Association of University Presses.

It is the policy of The Pennsylvania State University Press to
use acid-free paper. Publications on uncoated stock satisfy the
minimum requirements of American National Standard for
Information Sciences—Permanence of Paper for Printed
Library Material, ANSI z39.48–1992.

Contents

Acknowledgments

The realization of this edited collection required the support and commitment of a number of people and institutions whom we'd like to acknowledge. We'll start by expressing our thanks to our contributors for their patience and perseverance through the writing and editing process. Without their intellectual commitment to the idea that the intersections of rhetoric and mathematics are an important and underexplored space for scholarly work, this collection would never have materialized. In particular, we'd like to extend a special note of gratitude to Edward Schiappa, whose enthusiasm for the project in its earliest stages gave us the encouragement to push it forward.

In addition to our contributors, we would also like to acknowledge the institutional support we've received in the development of this collection. Like most scholarly works, this one has benefitted from the feedback its authors have received at conferences. One conference in particular, the Alta Conference on Argumentation, has been uniquely important to the development of this collection. In 2015 the selection committee for the conference elected to support two panels on rhetoric and mathematics, which allowed the majority of the contributors to this collection to present early versions of their chapters for critical feedback. It also afforded one of the editors a rare opportunity to meet face-to-face with the majority of the contributors to sketch out a general outline of the collection and consider how the chapters might fit together intellectually to create a larger whole. The collection has also benefitted from institutionally supported leaves granted to the editors by Carnegie Mellon University and Lewis and Clark College, which allowed us to complete the final stages of the project, including the writing of the introduction and chapter 1 and the editing of the complete manuscript.

Finally, we would like to thank the series editors, Leah Ceccarelli and Michael Bernard-Donals, for their support of this project, as well as Ryan Peterson, our acquisitions editor, who has shepherded us with kindness and competence through the submission and revision process.

Introduction

James Wynn and G. Mitchell Reyes

Mathematics has always influenced public culture. Prior to the late twentieth century, however, that influence was rarely felt on a daily basis at the level of the individual citizen. This is not to diminish mathematics' profound influence on everything from religion and music (Pythagoras, Pascal) to warfare and art (Archimedes, Da Vinci) to optics and physics (Newton, Leibniz) but merely to note that these influences could, to some degree, be separated from the everyday circulation of public discourse and the production of public culture. Only in the late twentieth and early twenty-first centuries has such a separation become increasingly difficult to sustain. We see it in the rise of mass media, statistics and probability, computational power, surveillance technologies, and—perhaps in the most immersive, communication-tracking technology thus far—the internet. These are just a few of the phenomena necessary for the emergence of an information society, in which those who can mathematically, algorithmically mine enormous data sets enjoy power heretofore unimagined. While mathematics and algorithms should not be conflated, the increasing cultural influence of mathematics via algorithms and other means has attracted the attention of an interdisciplinary audience of scholars, including rhetoricians.

With a few exceptions, however, scholars and practitioners of rhetoric and mathematics have for the last two millennia been content to let their fields coexist as parallel enterprises, even policing the boundaries between them to reinforce their incompatibility. In *Arguments in Rhetoric Against Quintilian*, for example, sixteenth-century logician and mathematician Peter Ramus wrote, "mathematicians deal with arithmetic and geometry, men of learning and wisdom, not rhetoricians" (683–84). Two centuries later rhetorical scholar George Campbell penned words with a similar sentiment in his *Philosophy of Rhetoric* (1776): "[Demonstration] is solely conversant about number and extension, and about those other qualities which are measured by these. . . . Here rhetoric it must be acknowledged has little to do" (65). Little changed during the two hundred years following Campbell's declaration of incommensurability, until in the 1980s mathematicians Philip Davis and Ruben Hersh (1987)—who had been invited to participate in a discussion about argumentation in different disciplines—took up the challenge of exploring the possible interrelations between rhetoric and mathematics. In their landmark contribution to the subject, "Rhetoric and Mathematics," they concluded "that mathematics is not really the antithesis of rhetoric, but rather that rhetoric may be sometimes mathematical, and that mathematics may sometimes be rhetorical" (54). Their work, supported by other intellectual undercurrents in rhetoric (discussed in chapter 1), encouraged a few rhetorical scholars to take up the study of mathematics, but these efforts were isolated, uncoordinated, and without designs to make the case for a sustained program of study. They also ran parallel to and without much acknowledgment of scholarship in the fields of history, sociology, philosophy, and mathematics education, each of which dealt with complementary topics. Despite a promising sea change in the perspectives of some researchers about the relationship between rhetoric and mathematics, no coordinated effort to study the intersections of these two fields emerged.

This book aspires to change this state of affairs. Though our chapters touch on myriad topics from a variety of perspectives, the volume is collectively dedicated to the argument that rhetorical scholars can and should make a sustained and coordinated effort to study the rhetorical dimensions of mathematics. For the last two centuries, mathematics has been woven ever more tightly into the social, political, scientific, and economic fabric of our lives. In James Wynn's book *Evolution by the Numbers* (2012), for example, he explored the process by which the study of evolution, variation, and heredity became mathematized starting in the mid-nineteenth century, a process that led to revolutionary new understandings and practices in breeding, medicine, and taxonomy even as it enabled the rise of new kinds of scientists such as biostatisticians (who conduct scientific research into biological phenomena by making it amenable to mathematical analysis). Similarly, scholars such as Theodore Porter (1995) have illustrated how mathematics became increasingly important in political decision-making starting in the nineteenth century because of its capacity to bridge the ethical gap between policy-makers and the nonexpert publics whose behaviors they wished to influence. More recently, the rise of digital technologies and increasing automation has made mathematics even more influential across a broad spectrum of political, economic, and sociocultural activities. Algorithms flag hate speech and monitor the influence of foreign powers in our political discourse and on our elections. They track the movements of people, money, and disease across the globe. They make forecasts about the future of our climate and estimate who will be the winners and losers in our future economy. As mathematics insinuates itself ever deeper into the social, political, scientific, and economic activities of our lives, the imperative grows for understanding precisely how it influences the values and beliefs with which we assess the world and make decisions, as well as how our values and beliefs influence the kinds of mathematical instruments we construct and accept. This interleaving both opens mathematics up to rhetorical analysis and makes it an important site for rhetorical studies.

The promise of a research space is typically judged by the number and quality of scholars occupying it and by the degree to which these scholars believe the concepts and methods native to their area of expertise can be used productively to produce new insight. To most scholars, rhetoric and mathematics may seem like largely uncharted territory. This impression is understandable given the current state of rhetoric and mathematics scholarship. Much of the work is distributed across academic space and time, appearing in different journals in different fields authored by researchers with few social or intellectual ties with one another. Thus the work of the first two chapters of this volume is primarily synthetic, bringing together the research of scholars interested in topics germane to rhetoric and mathematics and highlighting the intellectual connections between them in order to give some shape and coherence to the transdisciplinary and intradisciplinary conversations about rhetoric and mathematics that have already occurred. Chapter 1, for instance, explores the intellectual dissociation of rhetoric and mathematics that separated the fields for so long, teeing up our discussion later in the chapter on the transdisciplinary conversations that have begun to reassociate them. Building on these transdisciplinary conversations, Edward Schiappa's chapter, "In What Ways Shall We Describe Mathematics as Rhetorical?," examines the *intradisciplinary* discussions in rhetorical studies about mathematics and mathematical discourse. From his synthetic investigation a tripartite configuration emerges, with scholarship clustering around (1) rhetoric *of* mathematics, (2) rhetoric *in* mathematics, and (3) mathematical language *as* rhetorical. Schiappa uses this tripartite taxonomy to both organize and place into dialogue disparate work on the rhetorical dimensions of mathematics, leading ultimately to an examination of how rhetorical study of math allows scholars to see rhetoric itself in novel and unexpected ways.

While the opening two chapters collect and frame past scholarly conversations on rhetoric and mathematics, the body chapters deepen the exploration of questions such as "To what extent and in what ways does mathematics operate rhetorically" and "What insights can rhetorical scholars offer from its study?" We have organized these chapters into three clusters to help direct readers to particular emphases or interests. The first section deals primarily with the interface between rhetoric and math in public culture, the second with the interplay of rhetoric and math in moments of technical innovation, and the third with the synthesis of rhetoric and mathematics in contexts where experts seek to communicate persuasively to lay audiences.

Rhetoric, Mathematics, and Public Culture

Collectively, the first three body chapters address the power of mathematics to shape both public institutions and public culture. Both Cathy Chaput and Crystal Colombini's chapter and G. Mitchell Reyes's chapter examine how mathematics and mathematical ideas have influenced neoliberal economic orthodoxy and enabled massive economic bubbles like the one that led to the 2008 subprime crisis. Chaput and Colombini explore Adam Smith's metaphor of the "invisible hand" as part of the conceptual underpinnings of neoliberalism, arguing that while the persuasive force of the invisible hand is fundamentally rhetorical (i.e., it resides in the metaphor), this force is focused and directed by the mathematical constructs of neoliberal economics. They explain, "While this metaphor circulates *energeia*—a classical concept revived in recent rhetorical scholarship to discuss the intensity, power, and force that actualizes potentiality—its mathematical formulations crucially direct that power toward the kinds of economic activities that need to be cultivated in a given historical moment." In other words, the mathematics grounds the metaphor in real-world economic activities across shifting temporal contexts, making that metaphor more "real" in the process.

While Colombini and Chaput investigate the complex interrelationship between mathematics and metaphor, G. Mitchell Reyes explores how the commitments and assumptions of mathematical formulae, their "horizons of judgement," operate invisibly but tangibly to influence material practices of public culture. Using a combination of constitutive theory and Latourian actor-network theory, Reyes traces the rise of a little-known mathematical algorithm called the Li Gaussian copula, which played a crucial role in the growth and spread of subprime mortgages. His analysis unpacks the copula's horizon of judgment as well as the symbolic-material relations that it introduced into the domain of structured finance. These new symbolic-material relations, Reyes argues, fundamentally changed the size of structured finance and the power structures governing it, precipitating the collapse of the subprime mortgage market and the broader global economy in 2008.

Finally, Nathan Crick and Andrew Jones go beyond economics to show how mathematical ideas have shaped forensic investigations in both literature and the real world. Specifically, they trace Edgar Allan Poe's use of mathematically informed analytic logic and its influence on the public's imagination about the potential for a disciplined forensic approach to criminal investigations. Drawing on Peirce's divisions between rhetoric, logic, poetics, and mathematics, the authors describe how Poe's rational detective, Dupin, systematically develops a hypothesis for explaining the murders in the Rue Morgue. Their description highlights the role of mathematical reasoning as an inventional resource for building plausible stories of causality in literary plots, and they illustrate how Poe's detective story ignites the American public's interest in "rational forensic" methods. Jones and Crick conclude the chapter by exploring the real-world consequences of Poe's rational forensics as it was used to "solve" the real-life murder of New Yorker Mary Rogers.

Mathematical Argument and Rhetorical Invention

While it is unsurprising that rhetoric and mathematics comingle in economics and politics where numbers, beliefs, and values collide, it is perhaps unexpected to find rhetoric and math commingling in technical fields like physics. Yet such a commingling is precisely what Joseph Little identifies and traces in his study of Japanese physicist Hantaro Nagaoka's Saturnian analogy to explain spectral emission lines. Little shows that much of the *orthos logos* of Nagaoka's mathematics, the guiding "right reason" for his equations, was entailed by his initial conceptual commitments to a unique Saturnian analogy. His analysis also reveals that analogical mediations can operate natively (that is, without the presence of natural language) within the symbolic system of mathematics and that these operations can provide the rationale for establishing connections between mathematical concepts and procedures and natural real-world phenomena.

Whereas Little examines the intersections of rhetoric and mathematics in twentieth-century physics, Jeanne Fahnestock explores the gradual entanglement of mathematics and rhetoric in the development of scientific visuals from the sixteenth to the nineteenth century. In "The New Mathematical Arts of Argument: Naturalistic Images and Geometric Diagrams," she shows how Melanchthon—in his effort to create a syncretic art of argument derived from all fields of reasoning including mathematics and rhetoric—introduced a new type of definition argument in *Erotemata dialectices* (1547) that depended both on visualization and on encounters with material objects. This new modality of argument, influenced by the then improved ability to create and reproduce naturalistic images in woodcuts and engravings, was taken up aggressively in ensuing decades by natural philosophers, who increasingly used images to generate arguments that relied on the available conventions of geometrical depiction to manage their scrutiny. By examining the changing features of visual representation in science from the Enlightenment to the nineteenth century, Fahnestock illustrates the interconnectedness of mathematics, scientific visualization, and rhetorical invention.

Mathematical Presentations: Experts and Lay Audiences

While the previous chapters argue collectively that rhetoric and mathematics are deeply connected in both the public and technical spheres and that a rhetorical perspective provides a valuable and necessary vantage point for the study of mathematics, the chapters in this section illustrate the value a rhetorical perspective might have for mathematics professionals. One of the pressing challenges math professionals face is that American students struggle with mathematical proficiency and are rarely drawn into postsecondary programs or careers in mathematics. In *Raising Public Awareness of Mathematics* (2012), contributor Reinhard Laubenbacher of the Society of Applied and Industrial Mathematics describes the challenges facing the mathematics community this way:

> In the end, the imperative for the mathematical sciences community to raise awareness of mathematics among the general public is clear: we want

to recruit the next generation of mathematicians; we want our university administrators to value mathematics. . . . We want the general public and elected representatives to support mathematics and provide adequate funding for the agencies that promote our research; we want our funding agencies to view mathematics as the central enabling technology for much of scientific progress. . . . We want our K-12 educational system to train students adequately. (53)

The fundamental question Laubenbacher poses but does not ask directly is "How?" How can mathematicians and their institutions persuade the public to embrace mathematics while at the same time training them to succeed in the field?

Because the study of rhetoric is in part the study of persuasion, there might be something to gain for mathematics professionals from an examination of mathematics from a rhetorical perspective, if only to increase the effectiveness of their communication. Our final two chapters address these exact issues. James Wynn's chapter explores efforts by Danica McKellar, an actress with a bachelor of science in mathematics, to persuade middle school girls to identify with mathematical study and consider math careers. He also examines the challenges her female-centered accommodation strategies faced in public discussions. Assessments of McKellar's efforts and the public response to them provide useful information about audiences important to the mathematical community and the challenges that may need to be addressed in engaging them.

Extending Wynn's focus on audience, Michael Dreher assesses efforts by mathematics professionals to reach lay audiences within the context of changing K-12 math curricula. Dreher's chapter thus necessarily engages with issues surrounding the management of mathematical education, exploring the challenges math educators face when they try to persuade the public to adopt new and unfamiliar methodologies for teaching mathematics. Understanding these challenges from a socially informed and audience-centered perspective, Dreher argues, is a useful point of view from which to understand the many rhetorical exigencies associated with mathematics education and, as a result, imagine the kinds of solutions, rhetorical or otherwise, that might be devised to address those exigencies.

Collectively, the chapters in this volume seek to expand our understanding of the many ways rhetoric and math increasingly intersect in contemporary culture. These intersections have been studied independently for decades by scholars in mathematics, politics, philosophy, linguistics, history, sociology, education, and rhetoric. By identifying and connecting these mostly independent nodes of transdisciplinary scholarship, we hope to illuminate the contours of a transdisciplinary conversation in the making and offer inspiration to students, established scholars, and anyone inside or outside of rhetorical studies who might be interested in exploring the intersections between rhetoric and mathematics—intersections that are reshaping public culture in increasingly consequential ways and call in a rising chorus for a critical account.

References

Campbell, George. 1868. *The Philosophy of Rhetoric*. New York: Harper and Brothers. http://www.archive.org/details/philosophyofrhetoocampuoft.

Davis, Philip J., and Reuben Hersh. 1987. "Rhetoric in Mathematics." In *The Rhetoric of Human Sciences: Language and Argument in Scholarship and Public Affairs*, edited by John S. Nelson, Allan Megill, and Donald N. McCloskey, 53–68. Madison: University of Wisconsin Press.

Laubenbacher, Reinhard. 2012. "Mathematics in the Public Mind: The USA." In *Raising Public Awareness of Mathematics*, edited by Ehrhard Behrends, Nuno Crato, and José Francisco Rodrigues, 47–55. Berlin: Springer-Verlag. https://doi.org/10.1007/978-3-642-25710-0.

Melanchthon, Philip. 1547. *Erotemata dialectices: Continentia fere integram artem, ita scripta, ut iuventuti utiliter proponi possit*. Wittenberg: Lufft.

Porter, Theodore. 1995. *Trust in Numbers: The Pursuit of Objectivity in Science and Public Life*. Princeton: Princeton University Press.

Ramus, Peter. 2010 [1549]. *Arguments in Rhetoric Against Quintilian: Translation and Text of Peter Ramus's "Rhetoricae Distinctiones in Quintilian."* Translated by Carole Newlands. Edited by James J. Murphy. Carbondale: University of Illinois Press.

Wynn, James. 2012. *Evolution by Numbers: The Origins of Mathematical Argument in Biology*. Anderson, SC: Parlor Press.

Part 1

Framing the Intersections

1

From Division to Multiplication | Uncovering the Relationship Between Mathematics and Rhetoric Through Transdisciplinary Scholarship

James Wynn and G. Mitchell Reyes

In the introduction we asserted that for centuries mathematicians and rhetorical scholars have been content to let their fields reside in separate intellectual dimensions, and that more recently scholars of rhetoric and in other fields have begun to interrogate this separation. Though we offered a few quotes to support these claims, that evidence is insufficient for explaining how these two fields became dissociated and what manner of transdisciplinary efforts have since been made to reunite them. In this chapter we endeavor to amend these shortcomings by pursuing three relevant questions. First, how did rhetoric and mathematics become conceptually isolated from one another? Second, how has transdisciplinary scholarship reconnected them? And, third, how might rhetorical scholarship on math align with those conversations?

By addressing these questions, we reveal that a more complex story about the relationship between rhetoric and math exists beneath the surface of contemporary conventional thought. To introduce this complicated relationship, we first trace the historical opposition between the two fields—a division most notably established in Aristotle's influential theoretical work on argument. Then we will be in a position to see the slow emergence of counternarratives about this relationship—particularly in the late nineteenth and early twentieth centuries. These counternarratives come not just from rhetorical scholars but from mathematicians, philosophers, historians, sociologists, and scientists too. It is within this context that we will begin to see how past work on rhetoric and mathematics from rhetorical scholars aligns with the broader transdisciplinary conversation and, ultimately, how the chapters in this volume contribute to the study of the many ways that math and rhetoric intersect: intersections that seem to multiply with the contemporary rise of big data, the mathematization of increasingly diverse fields of power, and the spread of algorithmic culture.

Drawing Boundaries: Perspectives on Rhetoric and Mathematics
from Antiquity to the Nineteenth Century

Since the emergence of written treatises on rhetoric in the fourth century BCE, scholars have shared their perspectives on the place of both rhetoric and mathematics within systems of argumentation. In classical Greek rhetorical theory, mathematics was considered a science (episteme) with its own particular subject matter that was associated most closely with demonstrative argument. In his works on argumentation (*Topics*, *Rhetoric*, and *Posterior Analytics*), Aristotle identified three divisions within argumentation studies—rhetoric, dialectic, and demonstration.[1] Rhetoric and dialectic, he explained, deal with arguments that are probable and based on endoxa, or the shared opinion of wise "men of a given type" (*Rhetoric* 1.2, 1356a30, 1356b33). Mathematics, on the other hand, was associated with demonstrative reasoning, whose goal was to assemble necessary premises (premises that are undoubtedly true) to demonstrate didactically the truth of a particular conclusion (Barnes 1969, 138; Campbell 1868, 8). In the opening lines of *Posterior Analytics*, Aristotle highlights this feature of demonstration using the example of mathematics to epitomize it. He writes, "all communication of knowledge from teacher to pupil by way of reasoning presupposes some preexisting knowledge. . . . It is thus that the mathematical sciences are attained and every art also" (1.1). Though demonstration was not considered unique to the art of mathematics, mathematics was considered emblematic of demonstrative reasoning because it maintained irreducible axioms (true statements) and combined these to reason toward certain conclusions that should be self-evidently true or false for anyone exposed to them in speech or writing (*Posterior Analytics* 1.2). These conclusions must be true or false, admitting no degrees of probability. These features of demonstration contrast with rhetorical and dialectical argument, in which, as Aristotle explains, "there are few facts of the necessary type. . . . Most of the things about which we make decisions . . . present us with alternative possibilities" (*Rhetoric* 1.2, 1357a22). Because rhetoric deals with possibility rather than necessity, it has utility for persuasion in the courts and in political deliberations. Demonstration's stepwise reasoning from true axioms to broader truths, on the other hand, aids the pedagogue's effort to inspire the student's conviction in the conclusions of mathematics and other subjects that experts have already accepted as fact.

In addition to separating mathematics from rhetoric by making the former emblematic of demonstrative argument, Aristotle's treatises also denied the importance of style and delivery in mathematical demonstration, further establishing conceptual barriers between the two fields. Style—the capacity to choose language appropriate to the audience, subject, and occasion of argument—and delivery—the manner in which a speech is orally presented—were considered aids to persuasion because they overcame what Aristotle called the defects of the audience (emotion, ignorance, inattentiveness, etc.) that made arguing only from facts often ineffective for the orator. He concedes in the opening of book 3 of the *Rhetoric* that "the arts of language [i.e., style and delivery] cannot help having a small but real importance, whatever it is we have to expound to others: the way a thing is said does affect its intelligibility" (3.1, 1404a10). However, he follows this statement immediately with a caveat: "Not, however, so much importance as people think. All such arts are fanciful and meant to charm the hearer. Nobody uses fine language when teaching geometry" (3.1, 1404a12). Here, Aristotle rejects the importance of delivery and style to mathematical demonstration, thereby limiting the potential role for rhetoric in mathematics. Centuries later rhetorical scholars like George Campbell maintained these limitations and even extended them beyond style and delivery to include the canon of arrangement as well: "[Demonstration] is solely conversant about number and extension, and about those other qualities which are measured by these. . . . Here rhetoric, it must be acknowledged, hath little to do. Simplicity of diction and precision in arrangement, whence results perspicuity are . . . all the requisites" (1868, 65).[2]

In Aristotle's system of argument, there seems to be little overlap between rhetoric and mathematics. Rhetoric is associated with dialectic and probable reasoning, while mathematics is allied with demonstration and reasoning from necessary premises to self-evident conclusions. Rhetoric is also connected with style and eloquence, for which mathematical demonstration has little use because of its native propensity for perspicuity and resistance to fanciful thought and language. However, in the treatises of the Roman rhetoricians Cicero and Quintilian we encounter a slightly different vision of rhetoric that has a more interactive relationship with mathematics. In particular, both scholars share the opinion that the ideal orator should be versed in the art of rhetoric and schooled in the topics of knowledge on which they are asked to hold forth. Unlike Aristotle, they maintain that the power of eloquence gained by the study of rhetoric is essential to the success of persuasion in every area, including mathematical arguments. In *De oratore*, for example, Cicero argued that though some scholars believed that rhetoric's capacity to aid in persuasion or conviction was limited, every sort of argument—even ones involving mathematics—could be improved by the study of rhetoric. Cicero voices his perspective and defends it dialectically in his fictional exchange between Scaevola and Crassus. In the dialogue, Scaevola, the antagonist, warns Crassus that scholars in every field will accuse him of theft of property if he maintains that the study of rhetoric prepares the orator to speak compellingly on subjects in other fields. He uses mathematics as an example of an art for which the orator's training cannot prepare him:

> [SCAEVOLA:] As for the claim you made at the close of your speech . . . that whatever the topic under discussion, the orator could deal with it in complete fullness. . . . I should be at the head of a multitude who would either fight you with injunction, or summon you to make joint seizure by rule of court, for so wantonly making forcible entry upon other people's possessions. . . . The Academy would be at your heels, compelling you to deny in terms your own allegation. . . . I say nothing of the mathematicians . . . with whose arts this rhetorical faculty of yours is not in the remotest degree allied. (1.10.41–44)

In spite of Scaevola's warning, Crassus remains steadfast in his opinion that in every area of knowledge the art of rhetoric is essential to securing understanding, conviction, and persuasion: "[CRASSUS:] I rather think I shall come short of convincing you on my next point—at all events I will not hesitate to speak my mind: your natural science itself, *your mathematics*, and other studies which just now you reckoned as belonging peculiarly to the rest of the arts, do indeed pertain to the knowledge of their professors, yet if anyone should wish *by speaking* to put these same arts in their full light, it is to oratorical skill that he must run for help" (1.13.61–62; italics ours). The phrase "by speaking" is italicized here because, for both Cicero and Quintilian, the connection between thought and speech plays a central role in understanding the connection and contribution of rhetoric to knowledge and knowledge-making in a broad range of disciplines, including mathematics. In the second book of *Institutio Oratoria*, for example, Quintilian argues for the importance of speech to knowledge. He claims that reason cannot be fully realized as reason until it is communicated through speech, thereby making the study of language, style, and delivery (i.e., rhetoric) synergistic with and in part constitutive of reasoning in other disciplines. He explains, "Reason by itself would help us but little and would be far less evident in us, had we not the power to express our thoughts in speech; for it is the lack of this power rather than thought and understanding, which they [animals] do to a certain extent possess, that is the great defect in other living things" (2.16.13–15).

Drawing on this rationale—that speech is the means by which reason is expressed publicly—Quintilian, like Cicero, argues that "the material of rhetoric is to be composed of everything that can be placed before it as a subject for speech" (2.21.4). Though Quintilian recognizes that other disciplines have knowledge that cannot be learned through rhetorical study, he argues that because reasoning with and about that knowledge requires speaking, to speak well on a subject requires knowledge of rhetoric. It also follows that anyone possessing rhetorical training could learn about mathematics, music, farming, or architecture and, after having grasped the subject matter, speak on it with more eloquence and persuasive power than a mathematician, musician, farmer, or architect. Quintilian writes, "An orator should not be actually ignorant of the subject on which he has to speak. For he cannot have a knowledge of all causes, and yet he should be able to speak on all. On what then will he speak? On those which he has studied. . . . If the orator received instruction from the builder or the musician, he will put forward what he has best learned better than either" (2.21.16–17).

For Quintilian and Cicero, the boundaries between rhetoric and mathematics, or between rhetoric and any field of knowledge, for that matter, were predicated on subject matter or content. Because the subject matter of rhetoric brings shape to content so that it can achieve its object, the content in other areas *is* rhetoric's subject matter.[3] From this synergistic perspective, rhetoricians ply their craft on mathematical subject matter to enliven it and strengthen its persuasive or convictive capacity. Simultaneously, Quintilian and Cicero also argued that orators should learn mathematics and other subjects so that they might improve their ability to argue in legal and political disputations, contexts of argument native to rhetoric.[4] To make the point, Quintilian asks rhetorically, "Will he [the orator] not deal with measurements and figures? And yet we must admit that they form part of mathematics. For my part I hold that practically all subjects are under certain circumstances liable to come up for treatment by the orator" (2.21.19–20).

The synergistic relationship between rhetoric and other disciplines, including mathematics, though popular all the way through the Renaissance (Cifoletti
2006), was eventually attacked by Enlightenment scholars such as Peter Ramus
who sought to strictly bound the domain of rhetoric. Ramus used the relationship between rhetoric and mathematics discussed by Quintilian to illustrate
the absurdity and messiness of the system of education laid out by the Roman
orators and to call for strict divisions between rhetoric and other fields of
knowledge. In *Arguments in Rhetoric Against Quintilian*, Ramus writes, "Yet now
Quintilian follows Aristotle's and Cicero's confusion of dialectic and rhetoric.
Indeed, he makes it worse by fabrications of his own, and by including in his
teachings all the disputes concerning all the arts he had read or heard something
about—grammar, mathematics, philosophy, drama, wrestling, rhetoric. We
shall distinguish the art of rhetoric from the other arts, and make it a single one
of the liberal arts, not a confused mixture of all arts" (Ramus [1549] 2010, 80).
To distinguish rhetoric from all other arts, particularly dialectic, Ramus limited
the relationship between speech and reason and elevated the latter over the former. By his doing so, rhetoric lost all connection with operations of the mind
and reasoning, including the canons of invention, arrangement, and memory:
"Invention, arrangement, and memory belong to dialectic and only style and
delivery to rhetoric. . . . The orator, says Quintilian, cannot be perfected without
virtue, without grammar, without mathematics, and without philosophy. Therefore, one must define the nature of the orator from all these subjects. . . . I consider the subject matter of the arts to be distinct and separate. The whole of
dialectic concerns the mind and reason, whereas rhetoric and grammar concern
language and speech. Therefore, dialectic comprises, as proper to it, the arts of
invention, arrangement, and memory" (Ramus [1549] 2010, 104).

Ramus's dissociation of rhetoric from reasoning significantly restricted the
conceptual territory shared by rhetoric and mathematics. These restrictions
were partially undone by subsequent rhetorical scholars such as George Campbell, who drew dialectic and rhetoric back together under the category of "moral,"
or probable, reasoning. Despite the reunion of rhetoric with reason, however,
rhetoric was no closer to mathematics for Campbell than it had been for Aristotle. Demonstration, with which mathematics was associated, remained separated from moral reasoning on the ground that the former dealt with abstract
truths for which there could be no probable arguments while the latter was
"founded on the principles . . . from consciousness and common sense improved
by experience" and admitted multiple positions with degrees of validity (1868,

65–68). A broader perspective on the intersections between rhetoric and mathematics did not emerge until the late nineteenth and early twentieth centuries, and then only piecemeal in the works of rhetorical scholars as well as in other disciplinary scholarship, including mathematics, history, sociology, philosophy, education, and linguistics. The following section traces this emergent transdisciplinary conversation in the hope of giving some coherence to these efforts and, ultimately, showing how they align with scholarship in rhetorical studies.

Crossing Boundaries: Transdisciplinary Investigations into the Relationship Between Rhetoric and Mathematics in the Twentieth and Twenty-First Centuries

One of the more influential forces to break down the Enlightenment division between rhetoric and math came from within the field of mathematics itself. In the early decades of the twentieth century, Hilbert's axiomatic approach to math was much *en vogue*, a central tenet of which was that a fully complete system of mathematics was possible, one that would admit no quarter for probability or persuasion but only necessity and compulsion (see Hilbert 2004; Lakatos 1976; Van Kerkhove and Van Bendegem 2012). The Hilbert program, however, was significantly damaged by Gödel's now-famous incompleteness theorems of the 1930s. Those theorems showed in part that no complex system of mathematical thought could ever be completely closed, which meant that the great Hilbertian hope of using axiomatic logical proof as a foundation of absolute truth for math was at best a dead end and at worst a fundamental misunderstanding of the enterprise of mathematical innovation (Lakatos, 1976).

Gödel's work (and the work of many others who followed, such as Thurston 1994) opened conceptual space for a constructivist theory of mathematics, which gained traction both inside and outside the field. Scholars such as Ernest (1997) point to the end of the nineteenth and beginning of the twentieth centuries as an important era for the emergence of constructivism, one in which we see the proliferation and circulation of intellectual theories—from Durkheim's collective conscious to Peirce and Saussure's semiotics to Husserl's and Heidegger's phenomenology—that challenged Enlightenment theories of mind, Cartesian mind/body dualism, and representationalist theories of language. It was within this intellectual milieu that three important scholars—Chaïm Perelman, Lucie Olbrechts-Tyteca, and Kenneth Burke—helped revive rhetoric as a central framework for studying reasoning and argument, introducing perspectives on the relationship between rhetoric, mathematics, and language that significantly expanded the shared conceptual space between these fields.

Perelman and Olbrechts-Tyteca's *The New Rhetoric* is credited with reviving the scholarly tradition of rhetoric, which had been dormant for almost a century due in large part to the Ramusian spirit of the Enlightenment (Garsten 2006). In their resurrection of the field, the relationship between rhetoric and mathematics maintained some of the traditional boundaries defined by their predecessors; however, their treatise also introduced a novel perspective. Like previous rhetorical scholars, Perelman and Olbrechts-Tyteca separated demonstration/mathematics and dialectic/rhetoric into two distinct spheres of argument: the former associated with necessary, self-evidential reasoning and the latter with probable reasoning. They describe the association between mathematics and demonstrative analytical reasoning in the following passage: "The logician is indeed inspired by the Cartesian ideal and feels at ease only in studying those proofs which Aristotle styled analytic, since all other methods do not manifest the same characteristic of necessity. . . . Under the influence of mathematical logicians, logic has been limited to formal logic, that is to the study of proof used in mathematical sciences" (1971, 2). In their work on rhetoric, Perelman and Olbrechts-Tyteca protest the degree to which the study of reasoning and argument had been limited to demonstrative mathematical reasoning, and they argue for the revival of the study of probable argument, associated with dialectic and rhetoric, which developed "*in terms of an audience*" and relied on discursive techniques to "*induce or increase the mind's adherence to theses presented for its assent*" (4, 5).

Though in some sense the traditional borders between rhetorical and mathematical argument remained unviolated in Perelman and Olbrechts-Tyteca's general taxonomy of reasoning, the authors did identify an overlap between these forms of argument not previously acknowledged in rhetorical scholarship. In their divisions of rhetorical techniques, they identify as the first division "quasi-logical argument," arguments that "lay claim to a certain power of conviction, in the degree that they claim to be similar to the formal reasoning of logic or mathematics" (1971, 193). Arguments in this category seem to carry the force of mathematical reasoning and logic because they carry the form of it. These include arguments of definition, reciprocity, transitivity, comparison, probability, and part-whole, among others. While it is beyond the scope of this introduction to detail the different types of quasi-logical argument, their mention is significant because with them Perelman and Olbrechts-Tyteca open up a new space for exploring mathematics as a form of probable argument that influences public deliberation.

Whereas Perelman and Olbrechts-Tyteca's *New Rhetoric* provides important theoretical frameworks for identifying how math might operate rhetorically outside the technical sphere of mathematics as a suasive force (including in the public sphere), in Kenneth Burke's work we see the beginnings of a theoretical perspective for considering mathematics itself as fundamentally rhetorical. Burke's definition of rhetoric as *symbolic action* was essential in this regard. In the *Rhetoric of Motives*, Burke defines rhetoric as "the use of language as a symbolic means of inducing cooperation in beings that by nature respond to symbols" (1950, 43). Burke's framing of rhetoric as symbolic action accomplished through "symbolic means" expands the boundaries of rhetoric beyond natural language to embrace symbolic communication more broadly, including mathematical argument. As Burke explains in *Language as Symbolic Action*, "Besides . . . verbalization or talk, 'symbolicity' would also include all other human symbol systems such as *mathematics*, music, sculpting, painting, dance, architectural styles, and so on" (1966, 28; italics ours).

These two theoretical projects—Perelman and Olbrechts-Tyteca on one side and Burke on the other—form the intellectual foundations for the program of rhetorical inquiry that Schiappa maps out in more detail in the next chapter. In Perelman and Olbrechts-Tyteca's work, we see the possibility of engaging in what Schiappa labels the "rhetoric *of* mathematics" and in Burke's we find the foundations for both the "rhetoric *in* mathematics" and "mathematical language *as* rhetoric." Yet how do these projects align with broader transdisciplinary interests in math as a practice of meaning-making that has profound sociocultural impact? In addressing this question, we found that much of the transdisciplinary conversation aligns well with scholarship in rhetorical studies even in places where it pushes rhetoricians to consider the interface between rhetoric and math anew. To bring these alignments and moments of opportunity into focus we offer a basic taxonomy around which the transdisciplinary conversation orbits, with constellations of research emerging relative to three pervasive interests: (1) the force of quasi-logical mathematical argument within public and (nonmathematical) technical spheres, (2) the power of audience to shape mathematical ideas and practices, and (3) the use of mathematics as a sociosymbolic resource for promoting thought and action.

Quasi-Logical Argument in the Public and Technical Spheres

Since publication of *The New Rhetoric*, many transdisciplinary scholars have explored the concept of quasi-logical mathematical argument. Their research has investigated its use across a broad spectrum of disciplines and studied the breadth of its influence in driving scientific theory, encouraging metaphorical choices, and standing in for personal credibility (*ethos*). In *The Mismeasure of Man*, for instance, biologist and historian Stephen Jay Gould describes how belief in the power of mathematical reasoning buttressed fallacious perspectives on the relationship between biological determinism and intelligence in the nineteenth and twentieth centuries (1996, 21). He emphasizes the attraction that mathematical methods had as they were taken up by nineteenth-century researchers studying human variation, evolution, and heredity: "[A trend] swept through the human sciences—the allure of numbers, the faith that rigorous measurement could guarantee irrefutable precision. . . . Evolution and quantification formed an unholy alliance; in a sense, their union formed the first powerful theory of 'scientific' racism—if we define science . . . as any claim backed by copious numbers" (1996, 106).

The power of mathematics to dominate and shape persuasion in the biological sciences has also drawn the attention of rhetorical scholars such as Celeste Condit, whose book *The Meanings of the Gene* recognizes the power of mathematics to encourage particular choices of metaphor (1999, 14).[5] She makes the connection, for example, between the rise of statistical methods in the study of variation, evolution, and heredity in the late nineteenth and early twentieth centuries and the use of the "stock-breeding" metaphor. She writes, "Pro-eugenics discourse was conveyed in a vivid and sustained fashion through a stock-breeding metaphor that had recently gained increased credibility through the appropriation of new statistical and cellular research into biological heredity" (Condit 1999, 25). Collectively, Condit's and Gould's work provides insight into the power of quasi-logical mathematical argument both to drive scientific reasoning and to influence the rhetorical choices that scientists and scientific popularizers make when communicating with the public about scientific research.

In addition to the natural sciences, transdisciplinary scholarship in history and rhetoric has also endeavored to explore the force of quasi-logical mathematical argument in the social and political sciences. In his book *Trust in Numbers*, for example, historian Theodore Porter examines how numbers became a driving force in policy-making in Western democracies. He explains: "Since the rules for collecting and manipulating numbers are widely shared, they can easily be transported across oceans and continents and used to coordinate activities and settle disputes. Perhaps most crucially, reliance on numbers and quantitative manipulation minimizes the need for intimate knowledge and personal trust" (1995, ix). As Porter demonstrates in his book, mathematics, because of its perceived transparency and objectivity, began to replace personal trust as a persuasive force in policy-making. Rhetorical scholar Carolyn Miller, however, responds to Porter's work, suggesting that this replacement of *logos* for *ethos* can in fact be pushed too far and ultimately damage the relationship between publics and policy-makers. In "The Presumption of Expertise: The Role of *Ethos* in Risk Analysis," Miller explores the failures of the U.S. government's use of quantitative risk analysis to assuage public fears about nuclear plant safety in the wake of the Three Mile Island accident. She concludes that the government's reliance on quantitative risk analysis as a replacement for its *ethos* "sacrificed its claim to public trust, and the resulting political difficulties [were] entirely predictable from a rhetorical point of view" (Miller 2003, 202). In combination, Porter's and Miller's work interrogates the value and shortcomings of quasi-logical mathematical argument in policy-making.

Mathematics and Audiences in the Technical Spheres

The concept of quasi-logical argument offers a useful theoretical frame for highlighting the connections between the disparate transdisciplinary investigations of Gould, Condit, Porter, and Miller as well as their potential to elaborate and extend the scholarly conversation about the rhetorical force of mathematics, particularly the force of mathematical argument in disciplinary contexts outside of mathematics and in the public sphere. While research on quasi-logical argument has studied the general persuasive power of mathematics over audiences, other transdisciplinary research has examined the power of audiences over mathematical practices and the role rhetoric plays in persuading them to accept mathematical ideas. For example, problems of audience in mathematical argument have been a central focus in recent transdisciplinary work of historical scholars. In "Rigorous Discipline: Oliver Heaviside Versus the Mathematicians," historian Bruce Hunt investigates why an important paper of Heaviside's, an influential mathematician and physicist, failed to gain support when it was first introduced.[6] In his investigation, Hunt examines the extraordinary rejection of the third part of Heaviside's 1894 paper "On Operators in Physical

Mathematics" by the Royal Society (1991, 73).[7] To explain this rejection, Hunt turns to rhetorical theory, particularly to the significance of the audience: "By paying close attention to the rhetorical strategies that were successful in particular circumstances—and to those that failed—we can gain important insights into the aims and interests of the mathematicians of a given time and place. In particular, by examining how mathematicians use the rhetoric of 'rigor' to draw and redraw the boundary separating their own discipline from physics and engineering, we can better understand . . . why mathematicians responded as they did" (Hunt 1991, 75).

In his essay, Hunt offers a detailed historical assessment of the belief system of late nineteenth-century British mathematicians, concluding that for this audience "rigor" was established by a process of rational deduction wherein novel mathematical conclusions or applications were derived from established procedures and formulae (1991, 88). He points out that this perspective clashed with Heaviside's inductive or empirical methods of proving. Because Heaviside's methods of mathematical proof were different from those of his audience and institutionally threatening to them, his mathematical work, though groundbreaking, was rejected "to protect their [the mathematicians'] interests, and ultimately their autonomy" from what they saw as encroachment by the physical sciences (1991, 94).

Like Hunt's historical work, recent rhetorical scholarship has also focused on audience values and beliefs to explain the success or failure of mathematical arguments within technical intellectual communities. Zoltan Majdik, Carrie Anne Platt, and Mark Meister's paper "Calculating the Weather: Deductive Reasoning and Disciplinary Telos in Cleveland Abbe's Rhetorical Transformation of Meteorology" (2011), for example, investigates how Cleveland Abbe—an important figure in early twentieth-century meteorology—initiated a paradigm shift toward a mathematically informed program of meteorological research with the help of rhetoric. According to the authors, Abbe artfully prepared his audience to transition from a *qualitative* to a *quantitative* research regime by framing the transition as a progressive evolution. They explain: "through the simultaneous rhetorical imagination of and appeal to a disciplinary telos—a purpose for the discipline that exists independently of the traditions, assumptions, methods and practices of meteorology—he [Abbe] critiques, and portrays change in the discipline as a natural, inexorable progression" (Majdik, Platt, and Meister 2011, 77–78). The authors show how Abbe's capacity for reading and rhetorically adapting to his audience helped smooth the way for the methodological changes that mathematized the field of meteorology.

As these papers reveal, mathematical arguments operate in complex contexts in which real-world considerations about disciplinary values and boundaries require arguers to consider human audiences. The recognition that audiences pass judgment upon and, therefore, influence the character of mathematical argumentation also seems to animate another vein of transdisciplinary scholarship interested in the conventions of style and genre in mathematics. Historian Alex Csiszar's article "Stylizing Rigor; or, Why Mathematicians Write So Well," for instance, disregards Aristotle's dictum about fine language and geometry and explores the aesthetic principle of elegance (simplicity and clarity) in mathematical writing. Using writing handbooks published by the American Mathematical Society (AMS) and philosophical commentaries on mathematical proof and aesthetics, Csiszar identifies two important conventions of mathematical writing that promote simplicity and elegance. The first is the use of elliptical proof—a practice in which "routine"[8] computations are not comprehensively explained. Instead, the author "merely indicate[s] the starting point, describe[s] the procedure, and state[s] the outcome" (Csiszar 2003, 244). Csiszar connects this practice with the stylistic aesthetic of simplicity in mathematical reasoning: "The most perspicuous style is one that keeps things interesting by eliding the 'routine' statements" (2003, 245). The second convention, which might be described as "rational ordering," concerns the arrangement of information in scholarly mathematical papers. As Csiszar explains, arrangement in mathematical writing typically starts with the definition of terms, moves to the presentation of a theorem, or deductive conclusion, and ends with the description of proof (the mathematical steps by which the theorem was deduced) (2003, 247). This ordering ensures that the conclusion, or theorem, the arguer aims to prove is available to the reader as a context for understanding the proof which follows. This arrangement promotes the aesthetical quality of clarity and encourages the reader to associate the mathematical work with the disciplined procedures of rational deduction. Quoting Wolfgang Krull's "The Aesthetic Viewpoint in Mathematics" (1987), Csiszar explains: "The spare structure of mathematical texts is an important factor in validating the work that mathematicians do. . . . 'Mathematicians are not concerned merely with finding and proving theorems; they also want to arrange and assemble the theorems so that they appear not only correct but evident and compelling. Such a goal, I feel, is aesthetic'" (2003, 251).

Though the stylistic conventions of elliptic proof and rational ordering address the argumentative needs of mathematicians in intradisciplinary contexts, Csiszar argues that they create significant barriers for extradisciplinary and nonexpert audiences. Elliptical proof, for example, prohibits any but disciplinary experts from following the reasoning of mathematical research articles (RAs) and gives the impression that perhaps mathematical reasoning is not as rigorous and self-evident as mathematicians claim it to be (Csiszar 2003, 245). Similarly, rational ordering creates a sense among nonexperts that mathematicians are answering questions that nobody asked. As Csiszar explains, "the particular brand of perspicuity which is gained by placing a theorem (one's conclusions) at the beginning, is had at the price of obscuring one's methodology and the origins of the work" (2003, 248). As a result, he argues, "the arbitrary relationship between presentation and discovery, the research context of a problem and the questions that gave rise to it are often systematically suppressed in print" (2003, 248).

Whereas Csiszar's historical work explores stylistic preferences in mathematical writing and their impact on nonexpert audiences, recent rhetorical work on genre investigates the variability in mathematical argument written for expert audiences, challenging long-held beliefs, like the ones expressed by Campbell, that in mathematics "the mind is regularly, step by step, conducted forwards in the same track, . . . nothing left to be supplied, no one unnecessary word or idea introduced" (1868, 24). Over the last five years, rhetoric and composition scholars such as Heather Graves, Shahin Moghaddasi, and Azira Hashim (see also Kuteeva and McGrath [2015]) have collaboratively authored a series of papers exploring the general genre structures of mathematical research articles (RAs) (2013), their introductions (2014), and their use of appeals to exigence (Graves and Moghaddasi 2017). With this research, they bring to light a number of features that, they argue, distinguish mathematical RAs from RAs in other fields such as science, engineering, law, education, and linguistics. Surprisingly, one feature of mathematical articles is that they are less likely than scientific ones to follow the chronological move/step patterns described by Swales (1990, 140–66). This irregularity, they argue, challenges assumptions about the rigidity of mathematical argument. Mathematicians, they conclude, might be more responsive to the rhetorical demands of their audiences and contexts of argument than we have imagined (Graves, Moghaddasi, and Hashim 2014, 8).

Of all the dimensions of rhetoric *in* mathematics that have received scholarly attention—audience, style, argument, and arguer—the last element remains the least developed.[9] The limited transdisciplinary research that has been done on mathematical arguers, or rhetors, focuses primarily on the values and beliefs that individuals or groups of mathematicians identify themselves with and how these identifications are used as sources of invention for building individual and disciplinary *ethos*. In "The Professional Ethos of Mathematics," for example, French sociologist Bernard Zarca quantitatively analyzes the responses by over one thousand professional mathematicians and PhD students to a 2002–2003 online survey soliciting personal perspectives on a variety of topics, including the truth value of mathematics and publication practices. Using a test of statistical significance, Zarca identifies perspectives connected with particular groups within the mathematics community.[10] Based on his findings he concludes, for example, that "pure mathematicians place more importance on the intrinsic pleasure of their work, [while] applieds [i.e., applied mathematicians] place more importance on the collective aspects underlying their work," and that mathematicians that label themselves as "highly rigorous ... seem more attracted to the collective aspects of their work, placing more importance than the less rigorous mathematicians on the dissemination of knowledge acquired in the discipline" (Zarca 2011, 175–76).

As this brief survey of transdisciplinary work on the relationship between audiences and mathematics/mathematicians suggests, scholars in history, sociology, and rhetorical studies have begun to complicate what were before the twentieth century fairly rigid conceptual boundaries between mathematics and rhetoric. Though these investigations cannot be considered an intentional program of rhetorical inquiry, one can identify trends in the transdisciplinary scholarship that track with the interests of rhetorical scholars. By creating conceptual constellations of scholarly work, we can begin to appreciate the myriad directions taken and the contributions made to expanding our understanding of rhetoric, mathematics, and their cultural entanglement.

Mathematics as a Sociosymbolic Resource for Thought and Action

In some areas of scholarship, such as visual rhetoric, rhetorical scholars have enthusiastically responded to Burke's call to expand the field from natural language to the symbolic (Foss 2005, 141). In areas such as mathematics, the scholarly response has been more subdued.[11] Researchers in other fields, however, have found the study of mathematical discourse as a sociosymbolic resource both fascinating and illuminating. Much of this scholarship builds not on the work of Burke but on Peirce's mathematical semiotics, which introduced a theory of sign that accounted for meaning in various modes of representation outside of natural language. In creating a more inclusive theory, Peirce, like Burke, saw the interrelatedness of communicative activity across a broad spectrum of phenomena, which included natural language and mathematics: "It has never been in my power to study anything—mathematics, ethics, metaphysics, gravitation, thermodynamics, optics, chemistry, comparative anatomy, astronomy, psychology, phonetics, economics, the history of science, whist, men and women, wine, metrology—except as a study of semiotic" (1958, 408). Peirce's inclusive vision of representation and meaning-making invites, as Burke's does, a reframing of mathematics that opens new pathways for exploration by researchers in the humanities and social sciences. In this frame, mathematics, like natural language, becomes a sign system that evolves within sociohistorical contexts, includes subjectivity, and involves persuasion.

In "Towards a Semiotics of Mathematics," philosopher and mathematician Brian Rotman turns to semiotics "to articulate what mode of signifying . . . mathematical activity is; to explain how . . . mathematical imperatives are discharged; and to identify who or what semiotic agency issues and obeys these imperatives" (1988, 10). The semiotics of mathematics devised by Rotman attempts to reunite the abstract and concrete facets of mathematics and highlight the central role played by mathematical symbols in this relationship. He embodies this reunion in the figure of the mathematician as scribbler: "Our picture of the Mathematician is of a conscious—intentional, imagining—subject who creates a fictional self, the Agent, and fictional worlds within which this self acts. But such creation cannot, of course, be effected as pure thinking: signifieds are inseparable from signifiers: in order to create fictions, the Mathematician scribbles" (1988, 11). By implicating sign in mathematical invention and thought, Rotman's mathematical semiotics both advances the system Peirce developed and lays a theoretical foundation for connecting rhetoric and mathematics at the level of signification.

Using Rotman's semiotic philosophy as a jumping-off point, researchers have identified previously unrecognized spaces where rhetoric and mathematics intersect. Mathematics education specialist Paul Ernest, for example, goes beyond the individual thought experiment to examine the importance of signs as common ground for intersubjective exchanges about mathematics.[12] He argues, "A semiotic perspective ... transcends the traditional subjective-objective dichotomy. For signs are intersubjective, and thus provide both the basis for subjective meaning construction, as well as the basis for shared human knowledge" (2006, 68). As an educator, he focuses his attention in particular on how common knowledge of mathematical concepts and procedures is built up within mathematical discourse communities, including exchanges between communities of experts and novices: "The primary focus in a semiotic perspective is on communicative activity in mathematics utilizing signs. This involves both sign reception and comprehension via listening and reading, and sign production via speaking and writing or sketching" (2006, 69).

The emphasis on semiotics in both mathematical thinking and communication (speaking and writing) in Ernest's and Rotman's work revives and lends philosophical support to Quintilian's perspective that thought cannot be divorced from speech and, therefore, cannot be separated from rhetoric, even in the case of mathematics. In fact, Ernest explicitly echoes Quintilian's belief in a liberal arts education that synergistically combines training in rhetoric and writing in mathematics: "If the discourse of mathematics is no longer seen as purely logical ... but as having a contingent persuasive function varying with context, then the rhetoric of mathematics must be explicitly addressed and taught in the lecture hall and classroom" (1999, 79).

Whereas Ernest and Rotman challenge the traditional boundaries erected between rhetoric and mathematics by connecting mathematical thought to communication through signification, linguist Kay O'Halloran challenges received beliefs about the distinctness of natural and mathematical languages. By examining the semantic development of mathematics over time in a systematic functional linguistics (SFL) framework, O'Halloran shows that though mathematics and natural language have diverged to have distinct semantic affordances and characteristics, mathematics remains connected to and relies on natural language for meaning making. She explains: "[Mathematical] symbolism developed as a semiotic resource which evolved from [natural] language. . . . Despite the new functionality of mathematical symbolism, it nonetheless requires a surrounding linguistic co-text to contextualize the symbolic descriptions and procedures that take place. The dependence on the linguistic semiotic suggests that the [mathematical] symbolism did not develop a well-rounded functionality" (2005, 96–97).

The evolution of mathematics from natural language to a distinct symbolic system described by O'Halloran is traced in detail by historian Lynn Stallings in "A Brief History of Algebraic Notations." In her paper, Stallings explains that when algebra first emerged in the Babylonian and Egyptian cultures (ca. 1700 BCE) it "was written in words [i.e., natural language] without the use of mathematical symbols": a stage she and O'Halloran both describe as *rhetorical algebra* (Stallings 2000, 230; O'Halloran 2005, 96). This format, however, was less than ideal for mathematical description because problems and their solutions were cumbersome to write out and "room for ambiguity exist[ed] without the precision of . . . [a] standardized, efficient symbol system" (Stallings 2000, 230). The move toward a more efficient system of symbols took place in 250 CE when Diophantus of Alexander introduced *syncopated algebra,* which included abbreviations and notations. Diophantus substituted the abbreviation ΔY, for example, for the Greek *dunamis* (ΔYNAMIΣ), or power, to represent an unknown value squared (Stallings 2000, 231). It was not until the sixteenth and seventeenth centuries, however, that a compact notational system for operations (=, +, −, etc.) emerged and combined with abbreviated notations to constitute *symbolic algebra,* the semiotic system we recognize as algebra in the modern context. By tracing the transformation of mathematical symbolic systems over time, Stallings's and O'Halloran's work illustrates how algebra's "mathematical symbolism developed as a semantic resource with a grammar through which meaning was *unambiguously* coded in ways which involved *maximal economy* and *condensation*" while simultaneously revealing its debt to natural language semiotics (O'Halloran 2005, 97). The historical and linguistic evidence supplied by Stallings and O'Halloran extends and supports both Peirce's and Burke's idea that mathematics and natural language are social-symbolic resources that change with the needs and goals of the communities that use them. In other words, as Quintilian and Ernest remind us, they participate in the rhetorical.

Expanding Horizons

In this chapter we have traced the intellectual history that gave rise to the modern dissociation between rhetoric and math, putting in context the transdisciplinary effort to challenge that dissociation. Such historical contextualization ultimately enabled finer distinctions to develop within the transdisciplinary literature between, for example, scholarly interests in the impact of math on policy making and interests in how mathematization reshapes technical fields such as biology or meteorology. In the process, affinities also emerged between transdisciplinary scholarship and rhetorical research on math. We have organized these affinities around three pervasive interests: (1) the force of mathematical argument within public and nonmathematical technical spheres, (2) the power of audience to shape mathematical ideas and practices, and (3) the use of mathematics as a sociosymbolic resource for promoting thought and action.

This current volume stands on the shoulders of this broad-ranging transdisciplinary and rhetorical scholarship. Although a vibrant conversation exists across disciplines on the entanglements of rhetoric and mathematics, many questions remain. We need more scholars and more scholarship to help us understand the broad social and cultural implications of seeing the world and ourselves through an ever-denser array of mathematical relations. As the chapters in this collection demonstrate, rhetorical scholars are well positioned to contribute to this evolving conversation.

Notes

1. While Plato's work played a significant role in the separation of rhetoric and mathematics, we focus on Aristotle here because (1) only in his theoretical work does one find a developed theory of rhetoric that separates it from the demonstrative enterprise of ancient Greek geometry, and (2) Plato's influence has been addressed extensively by other scholars, sometimes to the exclusion of Aristotle's work (see for example Latour 1999, chap. 7; Nussbaum 2001; and Reyes 2014).

2. In the *Philosophy of Rhetoric*, Campbell argues that features of style such as the figures had a place in every kind of address excepting mathematics. He writes, "There is . . . one kind of address to the understanding, and only one, which . . . disdains all assistance whatever from fancy. . . . The address I mean is mathematical demonstration" (1868, 24).

3. At the end of book 2 of the *Institutio Oratoria*, Quintilian writes, "A few critics have raised the question as to what may be the *instrument* of oratory. My definition of an instrument is *that without which the material cannot be brought into the shape necessary for the effecting of our object*" (2.21.24).

4. On the subject, Cicero says in *De oratore*, "No man can be an orator complete in all points of merit, who has not attained a knowledge of all important subjects and arts. For it is from knowledge that oratory must derive its beauty and fullness, and unless there is such knowledge, well-grasped and comprehended by the speaker, there must be something empty and almost childish in the utterance" (1.6.20).

5. James Wynn's work (2007; 2009; 2012) builds in a variety of ways on Condit's analysis.

6. Among Heaviside's many achievements, he is perhaps best known for inventing operational calculus to solve the problem of measuring current across electrical circuits.

7. What makes this rejection extraordinary, as Hunt explains, is that Royal Society members, which included Heaviside, had the right to have their papers published without going through the peer-review process. It was, therefore, highly unusual for a member's paper to be rejected as Heaviside's was, and its rejection suggests a strong disposition against it amongst the mathematical membership in the society (Hunt 1991, 72–73).

8. I am using scare quotes here because the definition of what is considered routine is subjective and murky according to Csiszar and his sources. It depends on whether the audience and the rhetor agree that a calculation made by the latter was free of any "unexpected tricks" (Csiszar 2003, 244).

9. Reyes (2004) is an exception to this rule within rhetorical studies.

10. Specifically, a chi-squared test.

11. To our knowledge, Merriam Allen's (1990) "Words and Numbers: Mathematical Dimensions in Rhetoric" is the only such scholarly work that addresses this dimension of rhetoric and mathematics.

12. Like Ernest, Lakoff's and Núñez's (2000) cognitive metaphor approach engages mathematical discourse at the level of meaning-making; their deep commitment to cognitive science, however, seems to undermine what could be a more sophisticated theory of tropes (Reyes 2014).

References

Aristotle. 1901. *Posterior Analytics*. Translated by E. S. Bouchier. Oxford: Blackwell and Broadstreet.

———. 1952. *Topics*. Translated by W. A. Pickard. In *The Works of Aristotle*. Oxford: Clarendon Press. http://classics.mit.edu/Aristotle/topics.html.

———. 1984. *The Rhetoric*. Translated by Rhys Roberts. New York: McGraw-Hill.

Barnes, Jonathan. 1969. "Aristotle's Theory of Demonstration." *Phronesis* 14 (2): 123–52. https://www.jstor.org/stable/4181832.

Burke, Kenneth. 1950. *A Rhetoric of Motives*. New York: Prentice-Hall.

———. 1966. *Language as Symbolic Action*. Berkeley: University of California Press.

Campbell, George. 1868. *Philosophy of Rhetoric*. New York: Harper and Brothers. http://www.archive.org/details/philosophyofrhet00campuoft.

Cicero. 1996. *De oratore*. Translate by E. W. Sutton. Cambridge, MA: Harvard University Press.

Cifoletti, Giovana. 2006. "The Algebraic Art of Discourse." In *History of Science, History of Text*, edited by Karine Chemla, 123–35. Dordrecht: Springer.

Condit, Celeste. 1999. *The Meanings of the Gene: Public Debates About Human Heredity*. Madison: University of Wisconsin Press.

Csiszar, Alex. 2003. "Stylizing Rigor: Or, Why Mathematicians Write So Well." *Configurations* 11 (2): 239–68. https://doi.org/10.1353/con.2004.0018.

Ernest, Paul. 1997. *Social Constructivism as a Philosophy of Mathematics*. Albany: State University of New York Press.

———. 1999. "Forms of Knowledge in Mathematics and Mathematical Education: Philosophical and Historical Perspectives." *Educational Studies in Mathematics* 38:67–83. https://doi.org/10.1023/A:1003577024357.

———. 2006. "A Semiotic Perspective of Mathematical Activity: The Case of Number." *Educational Studies in Mathematics* 61:67–101. https://doi.org/10.1007/s10649-006-6423-7.

Foss, Sonja K. 2005. "Theory of Visual Rhetoric." In *Handbook of Visual Communication: Theory Methods and Media*, edited by Ken Smith, Sandra Moriarty, Gretchen Barbatsis, and Keith Kenney, 141–52. Mahwah, NJ: Lawrence Erlbaum.

Garsten, Bryan. 2006. *Saving Persuasion: A Defense of Rhetoric and Judgment*. Cambridge, MA: Harvard University Press.

Gould, Stephen J. 1996. *The Mismeasure of Man*. New York: W. W. Norton.

Graves, Heather, and Shahin Moghadassi. 2017. "'Since Hadwiger's Conjection . . . Is Still Open': Establishing a Niche for Research in Discrete Mathematics Research Article Introductions." *English for Specific Purposes* 45:69–85. http://dx.doi.org/10.1016/j.esp.2016.09.003.

Graves, Heather, Shahin Moghadassi, and Azirah Hashim. 2013. "Mathematics Is the Method: Exploring the Macro-organizational Structure of Research Articles in Mathematics." *Discourse Studies* 15 (4): 421–38. https://doi.org/10.1177/146144 5613482430.

———. 2014. "'Let G ¼ (V, E) Be a Graph': Turning the Abstract into the Tangible in Introductions in Mathematics Research Articles." *English for Specific Purposes* 36:1–11. http://dx.doi.org/10.1016/j.esp.2014.03.004.

Hilbert, David. 2004. *David Hilbert's Lectures on the Foundations of Geometry, 1891–1902*. Edited by Michael Hallett and Ulrich Majer. Berlin: Springer.

Hunt, Bruce. 1991. "Rigorous Discipline: Oliver Heaviside Versus the Mathematicians." In *The Literary Structure of Scientific Argument*, edited by Peter Dear, 72–96. Philadelphia: University of Pennsylvania Press.

Krull, Wolfgang. 1987. "The Aesthetic Viewpoint in Mathematics." Translated by Betty S. Waterhouse and William Waterhouse. *Mathematical Intelligencer* 9 (1): 48–52. https://www.gwern.net/docs/math/1987-krull.pdf.

Kuteeva, Maria, and Lisa McGrath. 2015. "The Theoretical Research Article as a Reflection of Disciplinary Practices: The Case of Pure Mathematics." *Applied Linguistics* 36 (2): 215–35. https://doi.org/10.1093/applin/amt042.

Lakatos, Imre. 1976. *Proofs and Refutations: The Logic of Mathematical Discovery*. Cambridge: Cambridge University Press.

Lakoff, George, and Rafael Núñez. 2000. *Where Mathematics Comes From: How the Embodied Mind Brings Mathematics into Being*. New York: Basic Books.

Latour, Bruno. 1999. *Pandora's Hope: Essays on the Reality of Science Studies*. Cambridge, MA: Harvard University Press.

Little, Joseph, and Maritza Branker. 2012. "Analogy in William Rowan Hamilton's New Algebra." *Technical Communication* 21:277–89.

Majdik, Zoltan, Carrie Anne Platt, and Mark Meister. 2011. "Calculating the Weather: Deductive Reasoning and Disciplinary *Telos* in Cleveland Abe's Rhetorical Transformation of Meteorology." *Quarterly Journal of Speech* 97 (1): 74–99. https://doi.org/10.1080/00335630.2010.539622.

Merriam, Allen. 1990. "Words and Numbers: Mathematical Dimensions of Rhetoric." *Southern Communication Journal* 55 (4): 337–54.

Miller, Carolyn. 2003. "The Presumption of Expertise: The Role of *Ethos* in Risk Analysis." *Configurations* 11 (2): 163–202. https://doi.org/10.1353/con.2004.0022.

Nussbaum, Martha. 2001. *The Fragility of Goodness: Luck and Ethics in Greek Tragedy and Philosophy*. Cambridge: Cambridge University Press.

O'Halloran, Kay L. 2005. *Mathematical Discourse: Language, Symbolism, and Visual Images*. New York: Continuum.

Peirce, Charles S. 1958. *Selected Writings: Values in a Universe of Chance*. Edited by Philip P. Wiener. New York: Dover. Google Books.

Perelman, Chaïm, and Lucie Olbrechts-Tyteca. 1971. *The New Rhetoric*. Notre Dame: University of Notre Dame Press.

Porter, Theodore. 1995. *Trust in Numbers: The Pursuit of Objectivity in Science and Public Life.* Princeton: Princeton University Press.

Quintilian. 1920. *Institutio Oratoria.* Translated by H. E. Butler. Cambridge, MA: Harvard University Press.

Ramus, Peter. 2010 [1549]. *Arguments in Rhetoric Against Quintilian: Translation and Text of Peter Ramus's "Rhetoricae Distinctiones in Quintilian."* Translated by Carole Newlands. Edited by James J. Murphy. Carbondale: University of Illinois Press.

Reyes, G. M. 2004. "The Rhetoric in Mathematics: Newton, Leibniz, the Calculus, and the Rhetorical Force of the Infinitesimal." *Quarterly Journal of Speech* 90:163–88.

———. 2014. "Stranger Relations: The Case for Rebuilding Commonplaces Between Rhetoric and Mathematics," *Rhetoric Society Quarterly* 44:470–91.

Rotman, Brian. 1988. "Towards a Semiotics of Mathematics." *Semiotica* 72:1–36.

———. 2000. *Mathematics as Sign: Writing, Imagining, Counting.* Stanford: Stanford University Press.

Stallings, Lynn. 2000. "A Brief History of Algebraic Notation." *School Science and Mathematics* 100 (2): 230–35. https://doi.org/10.1111/j.1949-8594.2000.tb17262.x.

Swales, John. 1990. *Genre Analysis: English in Academic and Research Settings.* Cambridge: Cambridge University Press.

Thurston, W. P. 1994. "On Proof and Progress in Mathematics." *Bulletin of the American Mathematical Society* 30:161–77.

Van Kerkhove, Bart, and Jean Paul Van Bendegem. 2012. "The Many Faces of Mathematical Constructivism." *Constructivist Foundations* 7 (2): 97–103.

Wynn, James. 2007. "Alone in the Garden: How Gregor Mendel's Inattention to Audience May Have Affected the Reception of His Theory of Inheritance in 'Experiments in Plant Hybridization.'" *Written Communication* 24 (1): 3–27.

———. 2009. "Arithmetic of the Species: Darwin and the Role of Mathematics in His Argumentation." *Rhetorica* 27 (1): 76–100.

———. 2012. *Evolution by Numbers: The Origins of Mathematical Argument in Biology.* Anderson, SC: Parlor Press.

Zarca, Bernard. 2011. "The Professional Ethos of Mathematicians." Translated by Amy Jacobs. In *An Annual English Selection,* supplement to *Revue Française de Sociologie* 52:153–86. https://www.jstor.org/stable/41336866.

2

In What Ways Shall We Describe Mathematics as Rhetorical?

Edward Schiappa

Let *rhetorical practice* be defined as the use of persuasive efforts to gain adherence, and let *rhetoric* be the discipline historically associated with the study and teaching of such practices.

If there is a domain of human inquiry that historically has been considered a rhetoric-free zone, it is mathematics (Reyes 2014). However, if one maintains that all human endeavors have a rhetorical dimension, then, as Philip J. Davis and Reuben Hersh claimed in their important essay "Rhetoric and Mathematics," "mathematics appears as the dragon which must be slain" (1987, 54). As G. Mitchell Reyes notes, we otherwise risk leaving intact a dichotomy between rhetoric and mathematics that "renders obscure a practice of writing/thinking that only seems to grow in influence each day" (2014, 471). Thus the rhetorical turn of the late twentieth and early twenty-first centuries includes various efforts to describe the creations of mathematicians as rhetorical. This essay provides a three-part synthesis of such efforts: (1) the rhetoric *of* mathematics, understood as the persuasive argumentative use of mathematics; (2) rhetoric *in* mathematics, understood as the argumentative modes of persuasion found in written proofs and arguments throughout the history of mathematics; and (3) mathematical language *as* rhetorical, a sociolinguistic approach to the language of mathematics, an approach I augment with recently published writings of Thomas S. Kuhn that describe deductive proofs as the "purest" of tropes.

To describe aspects of mathematics as rhetorical is not an attempt to reduce or diminish the status of mathematics in any way. As was argued some years ago, rhetoric scholars cannot "reduce" *anything*; we simply describe artifacts or practices as rhetorical (Schiappa et al. 2002). As Richard Rorty noted, a description is only as valuable as the uses to which it is put. Efforts to describe math as rhetorical, I believe, help us to understand mathematics as a profoundly social and historical activity. Whether a dose of rhetorical sensitivity would assist mathematicians, especially mathematicians in training, is a question to which I shall return.

Furthermore, to describe mathematics as a kind of rhetoric prompts us to consider carefully what we *mean* by "rhetoric" and "rhetorical," as I will argue in part 3 when discussing Thomas S. Kuhn's paper "Rhetoric as Liberation." Though Kuhn was not a rhetoric scholar, his insights about language, persuasion, and science offer an interesting counterpart to the notion that dubbing something "rhetoric" necessarily implies liberation. Successful rhetoric *constrains* people, and, as we will see, it is a good thing that it does.

The Rhetoric *of* Mathematics

The "rhetoric *of* mathematics" refers to the deployment of mathematics to enhance the credibility and persuasiveness of a particular argument. A mundane example might be the advertising claim that "four out of five dentists surveyed recommend sugarless gum for their patients who chew gum." Such a claim takes a standard argument form—argument by authority—and gives it added persuasive strength through an act of quantification. Often the ability of a speaker to describe phenomena in quantitative terms gives the audience the impression that the speaker is in touch with "reality" in a more concrete way (cf. McGee 1994).

Indeed, we can find evidence of the deployment of quantification and mathematics (broadly defined) throughout most of Western history to advance a variety of intellectual concerns (Crosby 1996). For Plato, geometry profoundly shaped his understanding of reality of ideal forms, and allegedly (but doubtfully) he inscribed over the door to his Academy "Let no one ignorant of geometry enter here" (Suzanne 2004). Anyone who has applied to college in the United States in the past half century knows that standardized testing has enshrined mathematics as evidence of intelligence, mental agility, and ability to succeed in college.

Historically, the maturity of a science "has come to be judged by the extent to which it is mathematical. First come astronomy, mechanics, and the rest of theoretical physics. Of the biological sciences, genetics is top dog, because it has theorems and calculations. Among the so-called social sciences, economics is the most mathematical and offers its practitioners the best job market, as well as the possibility of a Nobel Prize" (Davis and Hersh 1987, 53).

There are endless possible examples of the rhetoric *of* mathematics across history, but I want to describe a few, for reasons that will become apparent. Let's begin by turning to astronomy. The full title of Isaac Newton's *Principia* is *Philosophiae Naturalis Principia Mathematica*, or *Mathematical Principles of Natural Philosophy*. The reason the book is often heralded as so pivotal for the history of science is that it promised to extend the certainty achievable in mathematics to the study of nature. Of course, the notion that nature can be understood and described in the language of mathematics can be dated back to the ancient Greeks and to European scientists that predate Newton, notably Johannes Kepler, who had begun to produce a mathematical physics. Kepler's laws of planetary motion, for example, generated predictions that could be verified, which became the ultimate test for most scientific disciplines.

Perhaps the best-known example of the success of Newton's mathematical approach to physics is the discovery of the planet Neptune. The planet Uranus was discovered by William Herschel in 1781, and in the decades after its discovery its orbit was carefully observed and recorded. There were irregularities in its orbit that initially seemed inconsistent with what we now call Newton's theory of universal gravitation, which states that any two bodies attract each other with a force that is directly proportional to the product of their masses, and inversely proportional to the square of the distance between them:

$$F = G\,\frac{m_1 m_2}{r^2}$$

The irregularities in Uranus's orbit could be explained, however, by hypothesizing the existence of another, as yet unobserved, planet that was large enough to influence Uranus. What we now know as Neptune had, in fact, been observed two centuries earlier by Galileo in 1612 and 1613 and later by other astronomers, but in these previous observations it was believed that what was being observed was a star (a good example of the adage that "observation is theory-bound"). In the mid-1840s, two scientists attempted to calculate the location (and eventually the mass and orbit) of this hypothesized planet: John Couch Adams in England and Urbain Jean-Joseph Le Verrier in France. Adams completed his calculations earlier than Le Verrier, but they were not sufficiently accurate to guide British astronomers to find the planet. Le Verrier sent a letter to astronomer Johann

Gottfried Galle on August 31, 1846, that included his estimated mass, orbit, and location of Neptune. That letter was delivered to Galle at the Berlin Observatory on September 23. That same evening, after Galle had searched for less than an hour, Neptune was spotted and declared "discovered," less than a degree from the position Le Verrier predicted.

This story illustrates the power of Newtonian physics and the synthesis of empiricism and mathematics in science (most recently told in a lively fashion by Thomas Levenson [2015]) to promote new discoveries and knowledge. It is worth underscoring the fact that it is Le Verrier who is credited with the discover of Neptune, even though he was not the astronomer who first found it in the sky. He simply did the math. He, in fact, had a competitor in the Englishman Adams, but Le Verrier is typically given priority, in part because his math was better and yielded a more accurate prediction of where one would find Neptune in the night sky.

Accounts of other scientific disciplines embracing the rhetoric *of* mathematics are slowly emerging. James Wynn's *Evolution by the Numbers: The Origins of Mathematical Argument in Biology* (2012) is the first book-length project of which I am aware that draws upon the rhetorical tradition to provide a historical account of how a discipline was transformed by the rhetoric of mathematics. Wynn provides a detailed and compelling account of the persistent rhetorical campaign over decades by supporters of a mathematical approach to science to argue that the mathematics of statistics and probability is relevant and valuable to research in the areas of genetic variation, evolution, and heredity. By "rhetorical tradition" I mean to call attention to how Wynn's account deploys concepts such as ethos, topoi, tropes, narratives, audience, argument by analogy, Toulminian warrants, and so on to elucidate how scientists wishing to advance the role of mathematical argument in their respective disciplines made their case.

The result of Wynn's efforts is a rich history that illustrates the messiness, if you will, of mathematization. Far from a narrative of the inexorable march of scientific progress, the story is punctuated by stops and starts, contextualized by a host of historical influences and unique personalities. Wynn never lets us forget that scientists are rhetors addressing audiences both receptive and resistant, and that with hindsight we can understand the "good reasons" (in Walter Fisher's sense of the phrase) that led to the acceptance or rejection of individual biologists' claims. I concur completely with his conclusion that "the value of mathematics, both as a rhetorical strategy and as a feature that can be argued for rhetorically, suggests that the use of mathematics demands the attention" of rhetorical scholars (2012, 230).

Not all efforts to deploy mathematics rhetorically are successful. Infamously, professor of mathematics Neal Koblitz (1981) wrote an essay titled "Mathematics as Propaganda," criticizing what he saw as the irresponsible deployment of mathematics by scholars such as Paul Ehrlich and Samuel Huntington. As Wynn illustrates, the introduction of mathematics to academic disciplines faces resistance and ultimately is accepted only if it proves itself over time to help solve the problems facing such disciplines. And, indeed, sometimes the lure of mathematical argument can lead advocates into the realm of the absurd. For example, in an article titled "How Video Will Take over the World," James L. McQuivey (2008), arguing for the importance of visual images in the internet age, made the claim that "a video is worth 1.8 million words." The claim caught on, at least among those in the marketing industry, because it seemed to provide concrete justification for investing in video production and did so in a manner that could justify, mathematically, the value of such investments compared to text or static images (Harris 2015). Upon closer inspection, it turns out that the calculations were based on the unproven adage that "a picture is worth a thousand words"—itself a bald assertion originally popularized by advertisers trying to sell ads with images. Hence a sixty-second video, made up of thirty frames per second, has 1,800 "pictures" (Savage 2015). If each is worth a thousand words, then the video is worth 1.8 million words, or 3,600 pages of text. As silly as this statistic is, the fact that it has continued to circulate widely points to just how alluring mathematics can be as a source of persuasion.

I end this section on rhetoric *of* mathematics with a nod toward the emergence of competing rhetorics of Big Data. "Big Data" is a phrase that has its own interesting history, initially referring to data sets so large that traditional methods of data analysis are inadequate, but now referring to almost any large or complex data set—potentially including words and images in addition to numbers—resulting from computer-assisted collection and aggregation tools. Big Data has become the subject now of a plethora of articles, books, training workshops, conferences, business products, MOOCs, and so forth (let's not even get into the topic of data visualization). In 2008, editor-in-chief of *Wired* magazine Chris Anderson published "The End of Theory: The Data Deluge Makes the Scientific Method Obsolete." The long and short of Anderson's argument is that given enough data, one can make reliable predictions without understanding why the predicted outcomes occur. As Anderson (2008) put it, "This is a world where massive amounts of data and applied mathematics replace every other tool that might be brought to bear. Out with every theory of human behavior, from linguistics to sociology. Forget taxonomy, ontology, and psychology. Who knows why people do what they do? The point is they do it, and we can track and measure it with unprecedented fidelity. With enough data, the numbers speak for themselves."

Not surprisingly, there is a counterrhetoric that contends just the opposite— that numbers do *not* speak for themselves and that, as Lisa Gitelman's edited collection proclaims, *"Raw Data" Is an Oxymoron* (2013). The book addresses such issues as the ways that different kinds of data and different domains of inquiry are mutually defining, how data are products of social construction at each point in the processes of their collection and use, and conflicts over what can—or cannot—be reduced to data. Without question, the rubric Big Data will continue to be used persuasively, sometimes for good, sometimes not. Regardless, it exemplifies what I have described as the rhetoric *of* mathematics.

Rhetoric *in* Mathematics

Rhetoric *in* mathematics can be understood as the argumentative and stylistic modes of persuasion found in written proofs and arguments throughout the history of mathematics. I have argued previously that the rise of "Big Rhetoric" (the position that everything, or virtually everything, can be described as rhetorical) was driven by two primary theoretical arguments—epistemological and sociolinguistic rationales for viewing all human communication rhetorically (Schiappa 2001). In this section, I suggest that the epistemological rationale can help explain what we mean by rhetoric *in* mathematics. The epistemological rationale is fueled by the argument that philosophical criteria used to distinguish superior ways of knowing from rhetorical forms of knowledge have been critiqued persuasively.

Philosophers along with historians and sociologists of science in the mid-twentieth century unofficially teamed up to challenge these binaries. In some critiques, the first column (table 2.1) was described as requiring an unachievable sense of absoluteness or timelessness, leaving us only with the second column (Nelson, Megill, and McCloskey 1987). This argumentative move, often attributed to "postmodern" theorists (accurately or not), is responsible for some of the

Table 2.1   Epistemological differences between rhetoric and other branches of knowledge

Not Rhetoric (Science, Logic, Philosophy)	Rhetoric
Certainty	Probable
Knowledge/Epistêmê	Opinion/Doxa
Objective	Intersubjective
Reality	Appearance
Fact	Fiction
Literal	Figurative

most vehement opposition to Big Rhetoric (Cherwitz and Hikins 2000). In other critiques, such as by the physicist, historian, and philosopher of science Thomas S. Kuhn and the neopragmatist philosopher Richard Rorty, the columns are *historicized*; thus the difference between what might be deemed rhetorical or nonrhetorical is a reflection of historically situated consensus, which is a social-behavioral rather than a metaphysical or epistemological criterion. In a framework I have described elsewhere as Sophisticated Modernism (2002), the point is that we do have such things as "facts," but the humanist rhetorical scholar wants to historicize these to remind us that they are the product of human symbol-making and persuasive efforts that resulted in sufficient agreement to *count* as facts. All language use may be metaphoric at one level, as Lakoff and Johnson (1980) argued decades ago, but most of the time we have no difficulty distinguishing between "live" metaphors and language whose metaphoric origins have been forgotten and that we have "literalized." The Sophisticated Modernist reminds us that the distinction is social and behavioral, not metaphysical.

Turning to mathematics, George Lakoff and Rafael E. Núñez's majestic *Where Mathematics Comes From* (2000) is devoted to unpacking the metaphoric roots of mathematical thinking. They are convinced that "many of the confusions, enigmas, and seeming paradoxes of mathematics arise because conceptual metaphors that are part of mathematics are not recognized as metaphors but are taken as literal" (2000, 7). The process of persuading and socializing mathematicians into treating metaphors literally is very much a rhetorical process.

Davis and Hersh describe the rhetorical situation of mathematics in broad strokes. Most people do not think of mathematical proofs as rhetorical: "Somewhere behind each theorem which appears in the mathematical literature, there must stand a sequence of logical transformations moving from hypothesis to conclusion, absolutely comprehensible, certified as such by the authorities in the field, verifiable as such by even the novice, and accepted by the whole mathematical community. *This impression is absolutely false*" (1987, 61; italics added). For Davis and Hersh, mathematical proofs are *arguments*, and long before students are taught particular topics in school, those argumentative proofs have had to prove persuasive with human audiences, however specialized the relevant discourse community may be. Even in classical Greece, a place and time often lionized for producing the model of mathematical proofs, there was considerable disagreement over the scope, usefulness, and appropriate methods of mathematics (Lloyd 2012).

Note that I am subsuming the study of argumentation under the umbrella of rhetoric, which is a definitional move with which most rhetoric scholars agree and which has been enshrined in most rhetoric textbooks, but a move that you could refuse. Similarly, no branch of mathematics can get off the ground without acceptance of a large number of stipulated definitions; this is conspicuously true of axiomatic-deductive approaches to proofs, which are the approaches most often thought of as free of rhetoric. Social agreement defines the practice and pedagogy of mathematics at every step (Ernest 1999). Refuse the dominant orthodoxy and you face the consequences. A rhetorical perspective does not deny the value of mathematical argument; rather it emphasizes the inescapably social dimensions of creating a language of mathematics that is then deployed to induce agreement with historically situated proofs. Indeed, what *counts* as a "proof" is a historically and socially contingent judgment (Chemla 2012). As Davis and Hersh put it, "mathematics in real life is a form of social interaction where 'proof' is a complex of the formal and the informal, of calculations and casual comments, of convincing arguments and appeals to the imagination and the intuition" (1987, 68). They conclude, "our confidence in the correctness of our results is not absolute, nor is it fundamentally different in kind from our confidence in our judgments of the realities of ordinary daily life" (1987, 68). Indeed, a recent experiment found significant disparity among research-active mathematicians about the validity of a published proof from elementary calculus, leading the researchers to conclude the obvious: "There is not universal agreement among mathematicians regarding what constitutes a valid proof" (Inglis et al. 2013, 279).

Reflecting on the character of proofs and progress in mathematics, Field Medal winner William P. Thurston claims that "mathematical knowledge and understanding were embedded in the minds and in the social fabric of the community of people thinking about a particular topic," and that studying the historical development of specific proofs illustrates how "mathematics evolves by rather organic psychological and social processes" (1994, 8–9). Though Thurston does not use the *r*-word, it is clear from his account that he has what I am calling a rhetorical sensibility when he says it is "dramatically clear how much proofs depend on the audience. We prove things in a social context and address them to a certain audience" (1994, 15). To be sure, different mathematicians perform the role of "author" in quite different ways, and there is disagreement among mathematicians about what counts as good writing (Burton and Morgan 2000). As Annie Selden and John Selden (2014) have demonstrated, one can think of mathematical proofs as a rhetorical *genre* of discourse that must be learned in a particular historical and social context (see also Bahls et al. [2011]). We do not need to rely on Kurt Gödel's incompleteness theorem, or note the limitations of efforts to automate theory proofs, to insist on the inescapable role of social (hence rhetorical) interaction at every step in mathematics, including "pure" mathematics (Restivo 1990), for one cannot even start a proof without agreement and cooperation, and, as Thurston notes, openness to being convinced by others (1994, 9).

Alan Gross (1990, 213) claimed, "science is rhetorical, without remainder." There are two ways to make sense of such a phrase (in addition to noting the clever math metaphor at work). Some have understood the phrase as implying a form of ontological reductionism—everything and anything *is* rhetoric and nothing but rhetoric. Without question, that is not what Gross meant. Rather, he was advocating a form of methodological reductionism by which any and all artifacts or practices of scientists could, at least in theory, be described in rhetorical terms—that "all components of disciplinary discourse are within its explanatory compass" (1990, 49). This is a position he has since qualified (Gross 2016), but in the twenty-five-plus years since first publishing those words he and his coauthors have produced an impressive compendium of scholarship illustrating the usefulness of such a perspective. Most of that scholarship is historical. It is much easier to see the rhetorical (persuasion aimed at gaining adherence) either at the cutting edge (if one has the relevant expertise) or with the benefit of hindsight. Once a matter is considered settled and "beyond argument" by a scientific community, the controversy resists rhetorical analysis. As Gross now puts it, the apparent "rhetoricity" of science is "time sensitive" (2016).

Even if the goal of scientists and mathematicians is to achieve a consensus such that a particular knowledge-claim is considered beyond argument, it is precisely *how* such a consensus is gained that is of interest to rhetoric scholars. For example, a good deal of historical research challenges the notion typically drilled into schoolchildren that there is only one correct way to solve a math problem. Different styles of mathematical argument throughout history have been explained not only by individual personality differences, but also by ethnicity, nationality, and religion (Mancosu 2009; Chemla 2012). The more detailed and sensitive a historical approach one takes to mathematics, the greater the variety one finds in how mathematical proofs are presented over time, and there is no apparent reason not to describe such variety as a function of persuasive intentions and practices (Chemla 2012).

Scholars of rhetoric have barely scratched the surface of rhetorics *in* mathematics. A few decades ago, a first wave of articles proclaiming that science is rhetorical helped establish the rhetoric of science as a distinct subfield. But that subfield would not have survived if there had not been a second wave of work that carried out the difficult work of historical case studies. Scholars had to achieve a certain level of expertise in the scientific texts about which they wished to write, and only a handful of individuals (like Alan Gross, Jeanne Fahnestock, John Angus Campbell, and Leah Ceccarelli) have proven up to the challenge. My chapter is a contribution to a slow but steady first-wave effort, while work such as Wynn's is producing crucial (and difficult) second-wave scholarship.

An additional example of such second-wave historical scholarship is found in G. Mitchell Reyes's essay "The Rhetoric in Mathematics: Newton, Leibniz, the Calculus, and the Rhetorical Force of the Infinitesimal" (2004). The idea of the "infinitesimal" did not fit with mathematics up until that point in time and was at odds with Euclidean precepts. Infinitesimals "defied empirical testing" and "broke with accepted mathematical practice," hence, according to Reyes, rhetorical argument "was the sole champion of the infinitesimal" (2004, 163). Described by Newton and Leibniz as "evanescent" or "nascent," infinitesimals were paradoxical ("smaller than any positive quantity and yet larger than zero") yet vital to the success of calculus. Reyes tracks the criticisms made of this concept and analyzes Newton's and Leibniz's responses. Though infinitesimals were eventually replaced with other concepts, the controversy surrounding them is reminiscent of how Kuhn describes the heightened role of persuasion during scientific revolutions; indeed, mathematics, and for that matter science, underwent dramatic changes as a result of the introduction of the calculus.

Is it useful for students in introductory (K-12) mathematics classes to think of the rhetorical dimensions of mathematics? Probably not, even though it is clear that mathematical pedagogy is itself thoroughly rhetorical (Ernest 1999). If we replace the word "rhetoric" with *"argument,"* however, we find considerable recent interest in "mathematical argumentation" as a social and pedagogical practice. A growing number of scholars are embracing Andrew Aberdein's (2009, 1–2) description of mathematicians analyzing and assessing mathematical reasoning as a "species of argument" that, as a social practice, has more in common with "ordinary" or "informal" argument than a finished axiomatic-deductive proof (see also Aberdein and Dove [2013]). Such a focus on the argumentative practices of mathematics, some scholars believe, can improve the way mathematics is taught by enhancing the reasoning skills of student mathematicians (Carrascal 2015).

Is it useful for practicing mathematicians to think of their writings as rhetorical? It seems fair to say that the jury is still out on this question (Porter and Masingila 2000; Shield and Galbraith 1998). At MIT all mathematics majors are required to take two classes that are "communication intensive," and it is fair to say that an audience-oriented rhetorical perspective is promoted in such classes. Practicing mathematicians themselves are split. Ongoing research by MIT's Susan Ruff and Lisa Emerson of Massey University of New Zealand includes interviews with mathematicians (from graduate students to senior scholars) about their experiences, attitudes, and perceptions of writing in mathematics. One of the interview questions is whether math writing is persuasive. The research is still in its early stages, but many interviewees respond that a proof does not persuade—it is either true or it is not and the logic "speaks for itself"—a decidedly nonrhetorical attitude. Others reflect an attitude closer to William Thurston or Paul Ernest's position that "rhetorical form plays an essential part in the expression and acceptance of all mathematical knowledge" (1999, 75). Such varied responses are not surprising, given the uneven reputation persuasion and rhetoric still have in academics.

It might be useful to take a page from the history of economics to see what happens when the *r*-word is introduced to understand a discipline. In 1985, Deirdre N. McCloskey published the iconoclastic book *The Rhetoric of Economics*. McCloskey, along with John S. Nelson and Allan Megill, previously had organized an important symposium at the University of Iowa on the Rhetoric of the Human Sciences, which featured an impressive assortment of scholars, including Thomas S. Kuhn and Richard Rorty. McCloskey's book includes a critique of the influence of what we might call naïve modernism on economics and included such sections as "Rhetoric Is a Better Way to Understand Science,"

"Rhetoric Can Improve Economic Argument," and "Rhetoric Is Good for You." McCloskey was convinced that a dose of what we might call rhetorical sensibility would open up alternative approaches to economic argument, but it seems fair to say that that hoped-for result did not blossom. Economists cling no less tightly thirty years later to mathematical "existence proofs" and null-hypothesis tests of statistical significance. The prospects for the acceptance of a rhetorical sensibility seem better in economic history, which parallels, in certain ways, the reception of rhetoric of science scholarship.

Returning to Rorty's point that a description is only as useful as the purpose to which it is put, perhaps the conclusion here is that recognizing the argumentative and stylistic modes of persuasion found in written proofs and arguments is at least important and helpful to historians and sociologists of mathematics (Cifoletti 2006), and potentially of value in the communication training of professional mathematicians.

The Language of Mathematics *as* Rhetorical

The last topic I address is mathematical language. This approach finds its rationale in the sociolinguistic theory most clearly articulated in the works of Richard Weaver (1970) and Kenneth Burke (1950), that all symbol use is rhetorical. Because mathematics relies on symbol use, the creations of mathematicians—including formulae, calculations, and proofs—are no less rhetorical than those of poets. Weaver claims that "language is sermonic" in the sense that whenever we offer a description or label a phenomenon we are "preaching" a particular way of making sense of it (1970, 201–25). And, of course, Burke has two famous (or infamous) definitional statements in *Rhetoric of Motives*: first, that rhetoric is "the use of language as a symbolic means of inducing cooperation in beings that by nature respond to symbols" (1950, 43), and second, that "something of the rhetorical motive comes to lurk in every 'meaning,' however purely 'scientific' its pretensions. Wherever there is persuasion, there is rhetoric. And wherever there is 'meaning,' there is 'persuasion'" (1950, 172; cf. Merriam 1990). It is not difficult to insert the notion of mathematical symbols and expressions as a specialized form of symbol use that induces a particular way of understanding. Such a point is primarily philosophical, and likely to be of interest mainly during disruptions in the prevailing symbolic order of mathematics, or as a point of interest to those interested in the comparative history of mathematics (cf. Lakoff and Núñez 2000).

What I am describing as the sociolinguistic account of mathematical language *as* rhetorical can be clarified with examples. The first I described in *Defining Reality* (2003), namely, competing descriptions of the chopping down of a tree. The point in that book is that there is no neutral description possible. Whether we say a tree is being murdered, the land is being developed for a parking lot, or simply a tree is being cut down, all descriptions highlight certain aspects of the situation and neglect others (competing values, assigning agency and responsibility, etc.). The same is true if the phenomenon is described in the language of mathematics. Trigonometry and calculus might be useful languages to describe aspects of the falling tree for some purposes but useless for others. Similarly, an ocean wave might be described as it is in part of Pablo Neruda's poem "El gran océano" (in Eisner 2004, 102–3):

> La ola que desprendes,
> arco de identidad, pluma estrellada,
> cuando se despeñó fue solo espuma,
> y regresó a nacer sin consumirse.

> The wave that you let loose,
> arc of identity, exploded feather,
> when it was unleashed it was only foam,
> and without being wasted came back to be born.[1]

Or with a formula like this one:

$$\nu = \sqrt{\frac{g\lambda}{2\pi}\tanh\left(2\pi\frac{d}{\lambda}\right)}$$

Each is a distinct way of making sense of a phenomenon and encouraging us to understand that phenomenon in a particular fashion, wonderful for certain purposes and useless for others, and precisely what Burke had in mind with his provocative notion that meaning, itself, is persuasive.

Not only is mathematical language rhetorical when contrasted to symbol systems external to it, but, within the history or across different cultures, sociolinguistic differences can become visible. Probably most readers have been raised with a number system we call "Arabic," which is most often represented in base ten. Because we know how to "translate" numbers across systems, we do not typically think about such systems as different symbolic languages for mathematics. However, just as something is lost when translating words across languages—and the loss might be trivial (a matter of nuance) or profound—so too is there something lost when treating numbers as simple substitutes across languages. For example, 97 in English prose is ninety-seven, or 90 (plus) 7. In French, the word for 97 is *quatre-vingt-dix-sept*, which is four twenties (plus) ten (plus) seven. In the Danish counting system, which combines elements of base ten and base twenty, 97 is *Syvoghalvfems*, which stands for seven (plus) 4.5 20s. In base twelve systems, such as the Nimbia dialect of northern Nigeria's Gwandara language, 97 is *gume tager ni da*, or (12 × 8) plus one. In a base fifteen system, such as Huli (found in Papua New Guinea), 97 is *ngui waraga, ngui kane-gonaga karia*, or (15 × 6) + seventh object of the seventh fifteen. Again, the wider we look—either by looking back through history or across different linguistic cultures—the greater the variety of numerical ways of making sense of the world (Mazur 2014). As K. David Harrison notes in *When Languages Die* (2007, 167–200), different number systems and means of calculation provide important insights as to how humans think and cope with the practical demands that lead to culturally specific counting and calculating habits.

The historical evolution of various symbolic forms of expressing mathematical concepts is a further illustration of a sociolinguistic understanding of mathematics as rhetorical. Those unfamiliar with the history of mathematics may not realize that most of the symbols used today—not only = but also +, −, Σ, ÷, exponents, the use of letters for unknown quantities, and so on—were all created relatively recently (Mazur 2014). Euclid's influential *Elements* is a cornerstone of geometry, yet uses nary a symbol. Proposition 7 in modern expression would be $(a + b)^2 = a^2 + b^2 + 2ab$, but translated from Euclid's Greek to English it reads, "If a straight line be cut at random, the square on the whole is equal to the squares on the segments and twice the rectangle contained by the segments" (Mazur 2014, 87).

Different symbolic notations emerged in fits and starts across the globe. The hegemony of current symbols should not blind us to the fact that different symbols were introduced, were competed, and ultimately were chosen over time. For example, the equal sign originally was created by Robert Recorde in

The Whetstone of Witte in 1557 and drawn as two lengthy parallel lines ═══════════. As David Bodanis (2000, 25) points out, we could have grown up with

$$e \parallel mc^2$$
$$e \longrightarrow mc^2$$
$$e \text{ .æqus. } mc^2$$
$$e] [mc^2$$

Joseph Mazur argues that many mathematical symbols function much as words do, by making connections between experience and the unknown, and are used "to purposely transfer metaphorical thoughts capable of conveying meaning through analogy and resemblance" (2014, 220). He suggests that there may be an aesthetic dimension to the competition among mathematical symbols that unfolded through history: "Cultural predispositions of aesthetically persuasive symbols may play into our emotional appreciation of beauty, in mathematics as well as in poetry and art." Beauty in mathematics, he suggests, is in a large part "attributable to the illuminating efficiency of smart and tidy symbols" (2014, 220).

A subfield known as ethnomathematics is devoted to studying the cultural differences in the way mathematics is conceived, represented, and practiced (Ascher 1991). Such differences may be as simple as how number-words are constructed in different languages to contrasting forms of mathematical practices across cultures and over time. The point is that though different components of the language of mathematics, including the base-ten system, are presented to us in our education as "givens," historically they should be understood as rhetorical "takens," as socially agreed-upon symbolic systems that encourage and constrain us to understand the world in a particular way. That there is a dominant and widely shared mathematical language in advanced societies does not deny its rhetoricity but illustrates the relationship between rhetorical success and "constraint" in this context. Thomas S. Kuhn addressed this notion of constraint explicitly.

Kuhn wrote only once about the topic of rhetoric. He did so as a respondent to a panel at the 1984 Iowa symposium I referenced earlier. His paper was not published as part of the book project that emerged from the conference, and it only recently has been made available online (Kuhn 2014; Schiappa 2014). One of the papers to which he responded was an earlier version of the paper by Davis and Hersh that I cited previously. Kuhn seems to chide Davis and Hersh for not going far enough, since he believes they leave the impression that the ideal structure of mathematical proof is nonrhetorical. He suggests instead that we "explore the idea that mathematical proof is itself a piece of rhetorical apparatus, a trope" (2014, 8). He continues:

> Tropes are rhetorical figures which, by substitution, disclose unsuspected consequences of the meanings of the terms they deploy. Since Poincare, however, it has been standard to suggest that mathematical proof simply unpacks, again by substitution, unsuspected consequences tautologically present in the proof's premises. Can it simply be wrong to suggest that tautological substitutions are the limiting case, and thus perhaps the purest, of all tropes? Could the most coercive of all forms of argument perhaps turn out to be rhetoric at its most pristine? Discourse would then surely have its tyrannies, and one would not know how to be without them. (2014, 8–9)

Algebra, for example, cannot get off the ground without the rule of symmetry and various commutative rules. Long before we *apply* algebra we must understand the symbolic order that decrees $3x \cdot 5y = 15xy$. Kuhn's brief comments suggest that the traditional division between persuasion and coercion is not as clean as the discipline of rhetorical studies thinks. The language and reasoning patterns of mathematics must be acquired, and the learning process can be described as "persuasive socialization." When we use mathematical language, we accept the authority that such language has earned historically. Just as becoming a speaker of English obligates us to understand and accept the conjunctive function of the word "and," so too as math learners we are obligated to understand and accept the additive function of a plus sign. Elaborate formalizations similarly require one to submit to the prevailing symbolic order if one is to "speak" as a mathematician. Kuhn continues, "language cannot foster group solidarity or the evolution of belief through discourse except by simultaneously constraining the options open to the members of the group it binds" (2014, 4). Kuhn is referring to scientists in this particular passage, but I see no reason why his insights cannot also be applied to the discourse community known as mathematicians. He concludes, "Rhetorical consciousness will not banish coerciveness from discourse unless discipline and cogency are to be banished as well" (2014, 9–10).

Since the language of mathematics is human-made, we can still call it rhetorical, even if it is a different sort of rhetorical language than "Four score and seven years ago." We might distinguish between dynamic and static rhetoric in a fashion that parallels "live" and "dead" metaphors. When a symbol/concept like the number zero (rather than the placeholder) or the infinitesimal $(1/\infty)$ was introduced and contested, social acceptance was not assured, meaning was contested, and alternatives competed. But once accepted by the relevant discourse community, additions to mathematical language become definitive, and compulsory to those who wish to join that community.

Conclusion

I have set aside ongoing debates in the philosophy of mathematics on such issues as realism and antirealism along with competing philosophical schools of thought such as Platonism, psychologism, intuitionism, social constructivism, and so forth, though as one might imagine the congeniality of the label "rhetoric" would vary considerably from one school to the next. I write as a rhetoric scholar and pedagogue to encourage those willing to call into question the notion that mathematics stands alone as the one human endeavor free of rhetoric, while at the same time encourage somewhat more specificity in what we have in mind when using "rhetoric" and "mathematics" in the same sentence (see also Reyes [2014]). Accordingly, I have advocated in this essay that we distinguish among (1) the rhetoric *of* mathematics, understood as the persuasive argumentative use of mathematics, (2) rhetoric *in* mathematics, understood as the argumentative modes of persuasion found in written proofs and arguments throughout the history of mathematics, and (3) mathematical language *as* rhetorical, a sociolinguistic approach to the language of mathematics that recognizes the historical and cultural specificity of all symbol use.

The value of descriptions of mathematical texts and practices as rhetorical is a matter that cannot be decided a priori, but only through the continued good work of scholars younger and more mathematically inclined than this author.[2]

Notes

1. Pablo Neruda, excerpt from "The Great Ocean," translated by Stephen Kessler, from *The Essential Neruda: Selected Poems*, edited by Mark Eisner. Copyright © 2004 by Stephen

Kessler. Reprinted with the permission of The Permissions Company, LLC on behalf of City Lights Books. http://www.citylights.com.

2. My thanks to audiences at Carnegie Mellon University and the University of Pittsburgh for their feedback on an earlier draft of this essay, and to Allan Adams, Alan Gross, Elvira Caselunghe, and Mathias Møllebaek for their help.

References

Aberdein, Andrew. 2009. "Mathematics and Argumentation." *Foundations of Science* 14 (1): 1–8.

Aberdein, Andrew, and Ian J. Dove, eds. 2013. *The Argument of Mathematics*. Dordrecht: Springer.

Anderson, Chris. 2008. "The End of Theory: The Data Deluge Makes the Scientific Method Obsolete." *Wired*, June 23. http://www.wired.com/2008/06/pb-theory.

Ascher, Marcia.1991. *Ethnomathematics: A Multicultural View of Mathematical Ideas*. Pacific Grove: Brooks/Cole.

Bahls, Patrick, Amy Mecklenburg-Faenger, Meg Scott-Copses, and Chris Warnick. 2011. "Proofs and Persuasion: A Cross-Disciplinary Analysis of Math Students' Writing." *Across the Disciplines: A Journal of Language, Learning, and Academic Writing* 8 (1): 1–19. https://wac.colostate.edu/docs/atd/articles/bahlsetal2011.pdf.

Bodanis, David. 2000. $E = mc^2$. New York: Berkley.

Burke, Kenneth. 1950. *A Rhetoric of Motives*. New York: Prentice-Hall.

Burton, Leone, and Candia Morgan. 2000. "Mathematicians Writing." *Journal for Research in Mathematics Education* 31:429–53.

Carrascal, Begoña. 2015. "Proofs, Mathematical Practice and Argumentation." *Argumentation* 29:305–24.

Chemla, Karine, ed. 2012. *The History of Mathematical Proofs in Ancient Traditions*. Cambridge: Cambridge University Press.

Cherwitz, Richard A., and J. Hikins. 2000. "Climbing the Academic Ladder: A Critique of Provincialism in Contemporary Rhetoric." *Quarterly Journal of Speech* 86:375–85.

Cifoletti, Giovanna. 2006. "Mathematics and Rhetoric: Introduction." *Early Science and Medicine* 11:369–89.

Crosby, Alfred W. 1996. *The Measure of Reality: Quantification in Western Europe, 1250–1600*. Cambridge: Cambridge University Press.

Davis, P., and R. Hersh. 1987. "Rhetoric and Mathematics." In Nelson, Megill, and McCloskey, *Rhetoric of the Human Sciences*, 53–68.

Eisner, Mark, ed. 2004. *The Essential Neruda: Selected Poems*. San Francisco: City Lights Books.

Ernest, Paul. 1999. "What Is Social Constructivism in the Psychology of Mathematics Education?" *Philosophy of Mathematics Education Journal* 12:43–55.

Gitelman, Lisa, ed. 2013. *Raw Data Is an Oxymoron*. Cambridge, MA: MIT Press.

Gross, Alan G. 1990. *The Rhetoric of Science*. Cambridge, MA: Harvard University Press.

———. 2016. "The Limits of the Rhetorical Analysis of Science." *POROI: An Interdisciplinary Journal of Rhetorical Analysis and Invention* 12 (1): article 4. https://ir.uiowa.edu/poroi/vol12/iss1/4.

Harris, Bonnie. 2015. "6 Reasons Why You Need a Video Budget in 2016." Business 2 Community, 13 Oct. 2015. http://www.business2community.com/video-marketing

/6-reasons-why-you-need-a-video-budget-in-2016-01352407#FmwkLH4WUOfQ fpmo.97.

Harrison, K. David. 2007. *When Languages Die: The Extinction of the World's Languages and the Erosion of Human Knowledge*. New York: Oxford University Press.

Inglis, Matthew, Juan Pablo Mejia-Ramos, Keith Weber, and Lara Alcock. 2013. "On Mathematicians' Different Standards when Evaluating Elementary Proofs." *Topics in Cognitive Science* 5:270–82.

Koblitz, N. 1981. "Mathematics as Propaganda." In *Mathematics Tomorrow*, edited by L. A. Steen, 111–20. New York: Springer.

Kuhn, Thomas S. 2014 [1984]. "Rhetoric and Liberation." *POROI: An Interdisciplinary Journal of Rhetorical Analysis and Invention* 10 (2): article 3. https://doi.org/10.13008/2151-2957.1196.

Lakoff, George, and Mark Johnson. 1980. *Metaphors We Live By*. Chicago: University of Chicago Press.

Lakoff, George, and Rafael E. Núñez. 2000. *Where Mathematics Comes From*. New York: Basic Books.

Levenson, Thomas. 2015. *The Hunt for Vulcan . . . and How Albert Einstein Destroyed a Planet, Discovered Relativity, and Deciphered the Universe*. New York: Random House.

Lloyd, G. E. R. 2012. "The Pluralism of Greek 'Mathematics.'" In *The History of Mathematical Proofs in Ancient Traditions*, edited by Karine Chemla, 294–310. Cambridge: Cambridge University Press.

Mancosu, P. 2009. "Mathematical Style." *Stanford Encyclopedia of Philosophy*. July 2. https://plato.stanford.edu/entries/mathematical-style.

Mazur, Joseph. 2014. *Enlightening Symbols: A Short History of Mathematical Notation and Its Hidden Powers*. Princeton: Princeton University Press.

McCloskey, D. N. 1985. *The Rhetoric of Economics*. Madison: University of Wisconsin Press.

McGee, Timothy Christopher. 1994. "The Mathematics of Rhetoric: A Typology of Quantitative Figures." PhD diss., University of California, Berkeley.

McQuivey, James L. 2008. "How Video Will Take over the World." Forrester, June 17. https://www.forrester.com/How+Video+Will+Take+Over+The+World/fulltext/-/E-RES44199.

Merriam, Allen H. 1990. "Words and Numbers: Mathematical Dimensions of Rhetoric." *Southern Communication Journal* 55:337–54.

Nelson, John S., Allan Megill, and D. N. McCloskey. 1987. "Rhetoric of Inquiry." In Nelson, Megill, and McCloskey, *Rhetoric of the Human Sciences*, 3–18.

———, eds. *The Rhetoric of the Human Sciences: Language and Argument in Scholarship and Public Affairs*. Madison: University of Wisconsin Press.

Porter, Mary K., and Joanna O. Masingila. 2000. "Examining the Effects of Writing on Conceptual and Procedural Knowledge in Calculus." *Educational Studies in Mathematics* 42:165–77.

Restivo, Sal. 1990. "The Social Roots of Pure Mathematics." In *Theories of Science in Society*, edited by S. E. Cozzens and T. F. Gieryn, 120–43. Bloomington: University of Indiana Press.

Reyes, G. M. 2004. "The Rhetoric in Mathematics: Newton, Leibniz, the Calculus, and the Rhetorical Force of the Infinitesimal." *Quarterly Journal of Speech* 90:163–88.

———. 2014. "Stranger Relations: The Case for Rebuilding Commonplaces Between Rhetoric and Mathematics." *Rhetoric Society Quarterly* 44:470–91.

Savage, Jonathan. 2015. "Is 1 Minute of Video Really Worth 1.8 Million Words?" BoldContent, November 26. http://boldcontentvideo.com/2015/11/26/is-1-minute-of-video-really-worth-1-8-million-words.

Schiappa, Edward. 2001. "Second Thoughts on the Critiques of Big Rhetoric." *Philosophy and Rhetoric* 34:260–74.

———. 2002. "Sophisticated Modernism and the Continuing Importance of Argument Evaluation." In *Arguing Communication and Culture: Selected Papers from the 12th NCA/AFA Conference on Argumentation*, edited by G. T. Goodnight, 51–58. Washington, DC: National Communication Association.

———. 2003. *Defining Reality: Definitions and the Politics of Meaning.* Carbondale: Southern Illinois University Press.

———. 2014. "Thomas S. Kuhn and POROI, 1984." *POROI: An Interdisciplinary Journal of Rhetorical Analysis and Invention* 10 (2): article 3. https://doi.org/10.13008/2151 -2957.1198.

Schiappa, Edward, Robert L. Scott, Alan G. Gross, and Raymie McKerrow. 2002. "Rhetorical Studies as Reduction or Redescription? A Response to Cherwitz and Hikins." *Quarterly Journal of Speech* 88:112–20.

Selden, Annie, and John Selden. 2014. "The Genre of Proof." In *Mathematics and Mathematics Education: Searching for Common Ground*, edited by Michael N. Fried and Tommy Dreyfus, 248–51. New York: Springer.

Shield, Mal, and Peter Galbraith. 1998. "The Analysis of Student Expository Writing in Mathematics." *Educational Studies in Mathematics* 36:29–52.

Suzanne, Bernard. 2016. "Let No One Ignorant of Geometry Enter." Frequently Asked Questions About Plato. http://plato-dialogues.org/faq/faq009.htm.

Thurston, W. P. 1994. "On Proof and Progress in Mathematics." *Bulletin of the American Mathematical Society* 30:161–77.

Weaver, Richard. 1970. *Language Is Sermonic.* Baton Rouge: Louisiana State University Press.

Wynn, J. 2012. *Evolution by the Numbers: The Origins of Mathematical Argument in Biology.* Anderson, SC: Parlor Press.

Part 2

Rhetoric, Mathematics, and Public Culture

3

The Mathematization of the Invisible Hand | Rhetorical Energy and the Crafting of Economic Spontaneity

Catherine Chaput and Crystal Broch Colombini

The Chicago school of economics rose to fame by intellectually liberating capitalism from the encroachment of government intervention and allowing the market to self-organize in dynamic give-and-take with consumer desire. Proponents of this economic vision frequently assert that only an open market can enable the flawless mechanism of what Adam Smith called capitalism's invisible hand. Free from regulatory misdirection, individuals follow internal, self-interested desires that operate in lockstep with maximum utility, securing a stable economy and a thriving democratic state. Not merely conjectural, such propositions have been fodder for numerous mathematical proofs. Milton Friedman's positivist economics, developed in the 1950s and popularized in the 1980s with the best-selling *Free to Choose*, stands out among these. Coauthored with his wife Rose, the book begins by locating "Adam Smith's flash of genius" in the recognition that "economic order can emerge as an unintended consequence of the actions of many people, each seeking his own interest" (1990, 13–14). For Friedman, it matters little whether individuals *actually* make decisions this way. If economists can derive models—and governments can make policy—as if they do, the premise holds. By the 1990s this mathematically supported narrative became the basis for economic deregulation of industries worldwide. In 2008, however, the economy suddenly unraveled, leaving politicians and economists scrambling to account for the unfathomable: markets were free from government management and individuals were able to pursue their self-interest, but the result was a global financial disaster.

How did citizens and policy-makers become so thoroughly persuaded of something so improbable—that human action, insofar as it is wholly self-interested, will secure a market that is essentially consequence-free? This chapter

explores that question by inquiring into the rhetorical and mathematical constitution of the invisible hand. Mathematics is hardly a latecomer to rhetorical
conversations on economics. Indeed, it has become commonplace to note that
mathematical modeling persuades audiences of economists' ability to accurately
capture the dynamics of human economic behavior (Klamer and McCloskey
1998). Although many scholars further assert that such mathematical presentations obscure the sociopolitical content of economics and foreclose deliberation,
the relationship between these two epistemological fields strikes us as more
complicated. Mathematics has long figured in economic debates, including
those about capitalism's presumed equilibrium, without silencing either side.
David Ricardo represented the implicit mathematics of Smith's invisible hand
as a perfect equation, while Karl Marx's value theory opposed that idealization
using the same mathematical discourse. Keynesianism, the dominant economic
structure before the current neoliberal one, was deeply mathematical and also
invested in public deliberation. No doubt, mathematics has participated in the
rise of neoclassical economics to the exclusion of many heterodox perspectives,
but even a cursory glance suggests that scapegoating mathematics as a stylistic
feature that singularly quells further deliberation comes up short.

Just as it is reductive to indict mathematics for the lack of robust economic
discussion, it is equally problematic to concede to mathematics the lion's share
of the persuasive power wielded by economics. As we seek to show in this chapter, the mathematical logics that gird economic neoliberalism exist in a complex and evolving relationship with the doctrine's founding promise. For us,
that promise is concentrated in the enduring metaphor of Adam Smith's
invisible hand of the market and its empowering presumption that individuals, untethered by any obligation to consider the social scene, make the best
decisions based on their natural instincts. The process of participating in the
free market, in other words, is essentially effortless. While this metaphor circulates *energeia*—a classical concept revived in recent rhetorical scholarship to
discuss the intensity, power, and force that actualizes potentiality—its mathematical formulations crucially direct that power toward the kinds of economic
activities that need to be cultivated in a given historical moment. In short, we
contend that the mathematization of the invisible hand orients capitalist *energeia* so that particular economic behaviors become habituated as instinct.

To build support for this claim, we first survey the intersecting terrains
between the rhetoric of mathematics and the rhetoric of economics, arguing
that mathematics demands attention as a lever for capitalist *energeia*. Thus, we

trace the role of mathematics in the evolution of capitalist thinking. Our exploration begins with Adam Smith's famous invisible hand metaphor, moves through the mathematical economics of John Maynard Keynes, and ends with the popular *as if* doctrine of Milton Friedman. Whereas Keynesian economics used mathematical argument to correct and displace the invisible hand, Friedman's economics deployed mathematics to reanimate the energetic force of Smith's metaphor. These differing mathematizations of Smith's powerful and pithy metaphor do not represent objective discoveries that progressively refine collective economic knowledge; rather, they disclose the ability of economists to put mathematical methods to the politically motivated task of actualizing capitalist power toward concrete ends. Reversing the usual causation—mathematics valorizes economic formulations—we suggest that the invisible hand founds and energizes the mathematics of key economic thought, galvanizing an ongoing project of disproof and reproof. In each phase of capitalism, the prevailing mathematical presentation indicates the flow of economic work that must take place to ensure its promise of effortless abundance. Under the welfare state, that energy was funneled through government apparatuses; under neoliberalism, such energy moves through individual actors. Mathematical proofs determine the scope and location of appropriate economic activities. We illustrate this premise by exploring the financial literacy narrative that emerged from the 2008 economic crisis, stressing its role in realizing the rhetorical potential of Smith's invisible hand by cultivating the sophisticated financial subjects required of neoliberalism. We end with a discussion of the multifaceted relationships among rhetoric and mathematics as well as their importance to economic practices.

Mathematics, Economics, and the Constituting Power of *Energeia*

If disciplines exist along a perceived spectrum of rhetoricity, mathematics is typically thought to occupy a point at one far end where reality becomes self-referential. Consequently, when scholars explore mathematical and rhetorical intersections, the tendency has been to critique those who exploit mathematics as a neutral representation of worldly phenomena. Philip J. Davis and Reuben Hersh, for instance, conceive this relationship as a one-way street: sociopolitical analyses benefit from the rhetorical force of mathematics (1987, 53). They outline three kinds of mathematical use: pure mathematics with no real-world applications, applied mathematics that solves real-world problems, and rhetorical mathematics that adorns arguments with mathematics for suasive ends. For Davis and Hersh, Samuel Huntington's *Political Order in Changing Societies* (1968), which uses equations to explain geopolitical antagonisms, exemplifies rhetorical mathematics. Bereft of both mathematical insight and practical application, Huntington's equations do little more than harness the presumed rigor of mathematical reasoning, appropriating its disciplinary gravitas for partisan politics. This is the approach rhetorician James Wynn identifies, though with less sense of superfluity, in his study of how Charles Darwin used mathematics and its "ethos of correctness" to bolster his controversial evolutionary claims (2009, 92). In both cases, mathematical stylization is perceived to further authorial intentions through a reality-based force that propels arguments over the threshold of political suspicion.

Residing on the other end of the rhetoricity spectrum is the belief that mathematics is itself rhetorical. Representative of this position, G. Mitchell Reyes takes aim at the presumed authority of mathematics as a discipline above human dispute by illuminating the rhetorical nature of key mathematical concepts such as the infinitesimal. According to Reyes, the infinitesimal is less a mathematical element of the material world that was suddenly discovered by Isaac Newton and Gottfried Leibniz in the seventeenth century, and more a rhetorical invention forged to solve practical problems. By imagining the space under a curve as the summation of an infinite number of rectangles, mathematicians could measure more precisely the many dynamics of our physical world, even though the infinitesimal itself must be accepted on faith. The infinitesimal functions as "a rhetorically constituted concept" because it asks us to imagine a mathematical task that is empirically impossible (2004, 167). There is no mathematical or real-world proof of the infinitesimal—the litmus test demanded by Davis and Hersh for rhetoric-free mathematics—and yet it serves both disciplinary and applied functions. Undermining their tripartite division, the infinitesimal attests to "the constitutive rhetoric that emerges with the invention of novel mathematical concepts" (Reyes 2004, 181).

In the tug-of-war between these two rhetorical perspectives on mathematics, the view that mathematics lends others rhetorical credibility has been the victor to date. Its dominance is patently evident in the study of economics, where scholars regularly view mathematics as a reality effect rhetorically sutured to arguments in order to make uncertain claims certain. Indeed, theorists as politically divergent as free-marketer Deirdre McCloskey and democratic socialist

James Aune converge on one chord: mathematical encroachment into economics forecloses discussion. McCloskey's *The Rhetoric of Economics* argues that the econometrics of mathematical models obscures a wide range of rhetorical devices employed by economists. Such models, she argues, sometimes conceal dubious evidence and other times disguise sociopolitical conversations in the lifeless structure of numbers and facts. Extending discussion into the popular realm, Aune asserts that this "realist style" strives to place free-market economics—the ideological cry of neoliberalism—above deliberative debate. Despite stark differences in their approaches, McCloskey (1998) and Aune (2001) both believe economic discourse rhetorically avails itself of the perceived surety of mathematical logics.

Concern that the mathematization of economics replaces conjecture with certainty is neither new nor external to economics. As early as the 1930s, John Maynard Keynes critiqued the rise of what he called "symbolic pseudo-mathematical methods" for forcing assumptions of strict independence among determinants (1936, 297). His mathematical innovations took account of the social, psychological, and historical factors contributing to economic decision-making that remained exogenous to classical economic theory. As he saw it, those who used mathematics without incorporating these complicating realities were "blindly manipulating" variables and thus likely to lose sight of those "reserves and qualifications" essential to the fundamentally human science of economics (1936, 297). Keynes was not averse to formal modeling, he but maintained that models should be limited to those parts of economics that can be reasonably verified.[1] Other voices against the century's inexorable mathematical tide included economic historian Robert Heilbroner, whose popular 1953 book *The Worldly Philosophers* commemorated the theoretical traditions of classical economists and lamented the propensity of mathematical methods to erase normativity in economic judgments.[2] These critics do not reproach mathematical economics in toto, but only those varieties that claim distanced neutrality. They suggest that the increasing reliance on mathematical notation that characterizes contemporary economics cannot be appreciated fully without engaging the underlying values and operating assumptions of different models and their advanced equations.[3]

Breaking with rhetoric's tendency to engage economic mathematics only as a matter of style, we likewise wish to trouble the perception that quantification performs the bulk of contemporary economic persuasion. In nearly opposite fashion, we identify the rhetorical figure of the invisible hand as the backbone of economic orthodoxy and its compelling narrative. The invisible hand metaphor imbues capitalism's economic tenets and their mathematical representations with meaning through a constitutive intensity that promises a self-equalizing and undemanding market. Freed from any obligation to consider the needs of others, economic engagement becomes effortless as visceral desire drives economic decision-making and concerns about the social scene fade into the increasingly distant background. The vital energy or *energeia* moving through Smith's foundational metaphor as well as its relationship to both economic rhetoric and its mathematization require further investigation.

Foreshadowing subsequent interests in rhetorical *energeia* and its diverse orientations, George Kennedy began discussing energy as an underlying force common across different rhetorical forms and audiences in the 1990s. Perhaps best known for his translation of Aristotle, Kennedy shifted his attention from the Greco-Roman era to what he called a general theory of rhetoric. According to his theory, human beings have a "natural instinct to preserve and defend themselves," and rhetoric funnels that biological urge through language (Aristotle 1991, 7). Rhetorical energies assume different forms—"the emotional energy that impels the speaker to speak, the physical energy expended in the utterance, the energy level coded in the message, and the energy experienced by the recipient in decoding the message"—as they move throughout the communicative process (Kennedy 1992, 2). Crucial to his conception of a general rhetoric, this definition reappears almost word for word in Kennedy's later and more developed treatise, *Comparative Rhetoric*, which establishes energy as a common substance across rhetorical practices, including not just non-Western varieties but also animal and plant rhetorics. He only adds a short summative statement to his original proposition, noting that, as a consequence of energy, "rhetorical labor takes place" (Kennedy 1998, 5). An active force that transforms vibrant potential through concrete labor, rhetorical *energeia* pervades communication. As the physiological urge to defend, sustain, and thrive, rhetorical energy instructs our most taken-for-granted modes of engagement. Although such instincts might be a pervasive part of being in the world, they materialize according to the contingencies of their environment, a point he makes clear in his comparative accounts.

This enlarged definition of rhetoric intimates that the transfiguring force of *energeia* travels throughout the entirety of worldly existence to imbue all things, language only one among them, with the potency of life. Aristotle, for instance, discusses *energeia* across his scientific texts (*Physics, Metaphysics, De anima*) as the actualization of a power built into physical, abstract, and psychological things. Thus, his identification of rhetorical *energeia* as the process whereby language moves an audience into thinking and behaving differently describes one variation within a wider spectrum of energetic processes. In Monica Westin's reading of Aristotle, *energeia* enables the inherent power of numerous things to materialize, and rhetorical *energeia* names the specific process of "coming-into-being involving language" (2017, 258). As energy circulates through language and other material things, its actualizing power transcends the limits of its situatedness to include numerous other engagements. Most classical rhetoricians treated such things as beyond the purview of rhetoric. Aristotle, however, left an opening in his framework for this animating force by provocatively linking his concept of rhetorical *energeia* with other modes of material becoming, and others like Kennedy widened this opening by theorizing rhetoric outside of traditional speech acts.

Kennedy is not the only contemporary rhetorician to find inspiration in *energeia* as an active force that gives concepts mobilizing powers beyond mere referentiality. Ralph Cintron, for instance, argues that democracy has been brought to life—ontologized, he says—by the potency of social energies that move through its commonplaces to "organize our sentiments, beliefs, and actions" (2010, 100). As he sees it, democratic action is "deeply bound up with topoi produced from and furthering the *energeia* of collective passions and convictions" (2010, 102). Just as these energies maneuver through key signifiers such as democracy, they move through the individuals who employ them. Rhetorical energy or life vitality perpetually circulates and, in so doing, arranges the political, economic, and social patterns of historically situated being-in-the-world. The situatedness of a given speech act represents only an artificially isolatable moment of a dynamic process that continually constitutes and reconstitutes the terrain of reality through which persuasion takes place. As the world unfolds, energy has to be perpetually harnessed within new projects and directed toward different actions; it must be moved through specific symbolic orders to encourage and incentivize different behaviors.

For us, capitalism's most enduring metaphor—Adam Smith's invisible hand of the market—functions as a crucial rhetorical site in the ongoing movement of capitalist *energeia*. Analogous to Newton's and Leibniz's notion of the infinitesimal, the invisible hand is a rhetorical fabrication that nonetheless exerts concrete influence on the world. Through the many explicit and implicit iterations of this metaphor, citizens come to believe that the market will spontaneously follow the best course of action and self-correct when it missteps. Moreover, they understand themselves in a particular relationship to this infallible market: as individuals follow self-interest, all of society benefits. This constitution entails its own mathematical logic, implying that production equals consumption as long as economic behavior remains correlated to individual interests and desires. The rhetorical invention of the invisible hand calls mathematics into economic being, conscripting it within all future disputes over Smith's foundational claim. As we will show, the metaphor of the invisible hand regulates the energetic force of economic arguments—making them more and less powerful—while the mathematics directs that energy toward particular political economic practices. In short, the invisible hand metaphor harnesses rhetorical energy on behalf of capitalism generally, and its mathematical presentation funnels that energy toward capitalist particularities such as the cultivation of economic spontaneity.

In the following section, we illustrate this claim by following the ebb and flow of capitalist dynamics through different versions of the invisible hand.[4] We begin with Smith's initial formulation and then track it through two mathematical models: the refutation of John Maynard Keynes and the reclamation of Milton Friedman. Although the invisible hand metaphor never disappeared from the economic landscape, Keynes severely breached its omniscient authority, offering a rational economic schema through mathematical equations that described a precarious marketplace in need of government interventions to maintain long-term stability. Friedman overturned the Keynesian worldview by forging a mathematical schema consonant with Adam Smith's original construction of markets as guided by invisible forces. He ingeniously absorbed the primary *energeia* of free markets—the immensely compelling notion that they work without conscious individual effort—within the certainty of mathematical evidence. In arguments designed for public as well as disciplinary audiences, Friedman recharged capitalism through his *as if* doctrine and relocated appropriate economic intervention within individual subjectivity as opposed to governmental initiatives. In this way, economists have regularly relied on the invisible hand metaphor to compel individual actors to actualize the market's potentiality for equilibrium. The accompanying mathematization of the metaphor regulates, modulates, and directs those energies toward appropriate constituting activities. The invisible hand's contested mathematical terrain is not simply a way of retaining capitalism's origin story but a way of materializing its contemporaneous energies. An inherent paradox—the need to teach innate self-interest—will be the focus of the section after this, but we turn first to the historical evolution of the invisible hand.

Energizing Economic Mathematics: The Rhetorical Evolution of the Invisible Hand

Living in the first generation after Newton's and Leibniz's groundbreaking mathematical contributions, Adam Smith sought an all-encompassing principle to explain the connections among disparate economic phenomena. For Smith, all modern scientific accomplishments rest on the invention of a singular systemic power. As he saw it, such a unifying principle, "by representing the invisible *chains* which bind together all these disjointed objects, endeavours to introduce order into this chaos of jarring and discordant appearances, to allay this tumult of the imagination and to restore it, when it surveys the great revolutions of the universe, to that tone of tranquillity and composure" (Smith 1795, 20). Because these invisible chains exist apart from an individual's sensory capacity, they must be imagined rather than empirically derived. Once devised, however, unifying principles can be used schematically on behalf of concrete practices. Just as Newton's theory of universal gravitation revealed the force that governs the everyday behavior of physical objects, Smith aspired to disclose the force that governs ordinary economic behavior.

That force took shape in Smith's metaphor of the invisible hand of the market, a personification of the unifying power that enables people and institutions to work for the common good without any conscious intention to do so.[5] Smith described market economies as not beholden to the discrete actions of individual agents but organized by a transcendent power. According to his *Wealth of Nations*, an overarching sensibility inclines every individual to work toward the greatest revenue of the nation even though he "neither intends to promote the public interest, nor knows how much he is promoting it." On the contrary, "he intends only his own security; and by directing that industry in such a manner as its produce may be of the greatest value, he intends only his own gain, and he is in this, as in many other cases, led by an invisible hand to promote an end which was no part of his intention" (Smith 1991, 351–52). Although the explicit content of this passage is the risk-taking behaviors of merchants, Smith broadens this reference by using the subordinate clause to include "many other cases."

Making order out of the extensive chaos of exchange, the invisible hand is the connecting principle that explains the whole system. It guides individuals, groups, and nations; it ensures a morally, politically, and economically fair society; and it has all the information that fallible human beings do not. An adroit rhetorical figure, the invisible hand encourages people to follow their natural instincts without further thought about the causes or consequences of those inclinations.

Smith's powerful construction derives more from its metaphoric than its mathematical quality. This is not to say that *Wealth of Nations* is devoid of mathematical logic; on the contrary, Smith's market theory is often reasoned by way of such concepts as equilibrium and proportion. His case for improving the "productive powers of labor" relies on assertions of various proportionate relationships, including an inverse one between the profits of stock and the wages of labor (1991, 93). Still, it develops through many sections of prose that, lacking equations and other mathematical illustrations, invest Smith's rhetorically constituted principle of economic order with the collective energies of an entire society all working according to their own individual best interests. Smith's primary appeal derives not from mathematics but from his invisible hand metaphor and the way it brings the promise of effortlessness to life.[6] For Aristotle, rhetorical *energeia* reflects the ability to make "the lifeless living through metaphor" (Aristotle 1991, 249). It has an instructive quality inasmuch as it reveals, often by way of contrast, something that the audience did not previously understand. The invisible hand of the market exemplifies such criteria—it is a concise metaphor that works by contrast (a guiding hand that no one sees or feels) to signify activity (organizing optimal market relationships) and to offer instruction (pursue individual economic interests and society will take care of itself). Cultivating a powerful argument—society does not require difficult and time-consuming deliberations over economic action when everyone focuses on his or her individual goals—for the rejection of government intervention, Smith establishes the invisible hand to orient human activities.

It was not until John Maynard Keynes, a century and a half later, replaced the transcendent invisible hand with rational assessment and conscious intervention that mathematics became central to economic conversations. In accordance with his invisible hand theory, Smith maintained that capitalism naturally produces an equilibrium between supply and demand—everything manufactured by producers is absorbed by consumers. Keynes's most famous text, *The General Theory of Employment, Interest, and Money*, took this assumption to task by offering an alternative based on empirical experience and mathematical assessment. His theory of consumption stabs at the heart of the invisible hand by asserting that demand does not naturally regulate supply and, therefore, requires government intervention. To make this argument, he divides consumption into two contributing categories—autonomous factors such as food and shelter that are needed regardless of one's income and wage-dependent factors that become less certain as one's wealth increases.[7] Autonomous consumption, or the average cost of living, can be calculated fairly easily through price indices, a new economic tool of the twentieth century. However, the propensity to consume based on income is trickier to determine as it depends on a host of cultural and individual preferences that defy precise mathematical calculations. Keynes solves this dilemma with the premise that as income increases, a proportionally larger share of it goes into savings. At a certain point, income outpaces the possibility of luxury goods and people funnel surplus into savings. For Keynes, these additional savings represent money that is not reinvested in the economy. As a society's wealth increases, it becomes susceptible to a lack of investment that cannot be stimulated through monetary policy. Even with low interest rates, for instance, individuals may choose not to invest, believing that when rates go up, their money will be locked into less remunerative obligations. Keynes concludes that governments must intervene to maintain equilibrium and the mathematics that support this argument direct the invisible hand's social energies toward that regulation.

Keynes backs his theory of a propensity toward saving with the probabilistic nature of economic reasoning, a justification that does not fully dispense with Smith's invisible hand but does undermine its implicit equilibrium. Because probability focuses on "arguments [that] are rational and claim with some weight without pretending to be certain," it provides, he believes, a useful mathematical model for assessing the relationship among the many imprecise factors that determine economic decisions (Keynes 1921, 3). These decisions rely on diverse forms of evidence—direct experience, inherited knowledge, and instinctual ideas, among others. Not all are equally valid, however. Experience, he says, provides a rational basis for probable expectations and thus strengthens arguments. Conversely, instinct provides irrational evidence and weakens arguments because it lacks empirical support (Keynes 1921, 15). By demanding that mathematical inventions be probable, if not empirically certain, Keynes recasts Smith's invisible hand from an a priori premise to an inductive claim that can be assessed within the given circumstances. Transforming the invisible hand into an element of probabilistic reasoning, Keynes emptied Smith's rhetorical invention of its effortless certainty and established economic deliberation and government intervention as necessary to the maintenance of effective market functions.

Whereas Smith's invisible hand formulation enthralls with its effortless *energeia*, Keynes's reformulation clearly makes no such promise. Deliberative discussion and probabilistic calculation require more time and energy from citizen-subjects than do intuitive acts of self-interest. A difficult sell, Keynesian interventionism likely gained traction because of the extreme precarity of its economic context and was displaced as the dominant economic framework by the ascendancy of Milton Friedman's positivist mathematical economics during a similarly difficult economic scene—the stagnant wages and rising inflation of the 1970s. While Keynesian mathematics attempts to represent life's dynamic experiences so as to indicate the most probable course of action, positivist economics reverses this logic. Friedman advocated that economic theory mirror "the physical sciences" by pursuing positivist, rather than probable, mathematics (1953, 4). For Friedman, mathematical economics need not represent a realistic picture of life experience: "The relevant question to ask about the 'assumptions' of a theory is not whether they are descriptively 'realistic,' for they never are, but whether they are sufficiently good approximations for the purpose in hand" (15). Friedman spent his entire career creating an alternative economic lexicon that would support what he believed to be the main purpose at hand—legitimating laissez-faire economics with sound mathematical evidence. A heady rhetorical challenge, this required him to reconstitute key elements of economic equations so that he could present Smith's invisible hand theory through the scientific veneer of irrefutable mathematical conclusions. Much has been made of his mathematical scientism, but we wish to stress the importance of its energetic resonance with the invisible hand.

Central to Friedman's counter-Keynesian mathematics is his famous doctrine of *as if* by rational choice, which reinserts instinctual processes of social energy as foundational to economic theory. His methodology works from the premise that unconscious behavior efficiently performs what rational calculation inefficiently approximates. Friedman offers two examples to support this claim. First, he declares that a leaf positions itself "as if it knew the physical laws determining the amount of sunlight that would be received in various positions" even though leaves clearly "do not 'deliberate' or consciously 'seek,' [and] have not been to school and learned the relevant laws of science" (19–20). Additionally, he says, a successful billiard player executes "shots *as if* he knows the complicated mathematical formulas" (21). In these and other situations, "the result achieved by purely passive adaptation to external circumstances is the same as the result that would be achieved by deliberate accommodation to them." From this claim he derives the proposition that "individual[s] [and] firms behave *as if* they were seeking to rationally maximize their expected returns" because they intuitively conform to the structures around them and not because of rational decision-making. Neither individuals nor businesses, he says, "solve the system of simultaneous equations in terms of which the mathematical economist finds it convenient to express [a] hypothesis, any more than leaves or billiard players go through complicated mathematical formulas" (20–22). Asserting that economic agents intuit patterns of behavior that require mathematical gymnastics for economists to replicate, rational choice doctrine reclaims Smith's theory that unthinking behavior automatically aligns with the needs of capitalism, and thus the greater social good, precluding the need for conscious assessment and collective design.

Disputing Keynesian probability, Friedman uses positivism and mathematics to reshape the entire field of economics. Positivism authorizes him to revise the Keynesian consumption theory by using abstract numbers rather than concrete ones. He reconstitutes income from a defined annual salary to an average salary invented from a lifetime of presumed earnings. As a rule, Friedman argues, people do not consume based on an income at a given time but on income expected over a lifetime—what he calls "permanent income" (1957, 28). With this new definition of income, he concludes, "the ratio of savings to income is independent of the size of permanent income" (26). Through this rhetorically imagined income he simultaneously disproves the Keynesian assertion that increased wealth leads to increased savings and undermines its justification for government interventionism. Friedman deploys similar rhetorical inventions to develop his hallmark theory of monetarism. In order to prove a stable monetary demand (the amount of national savings), he adjusts the number plugged into the formula from an actual dollar amount—reported by banks, on accounting ledgers, and in tax returns—to an imagined quantity that he strategically names "real dollars" (1956, 10). These so-called real dollars represent what actual dollars, in general and on average, can purchase (21). If the national data are adjusted for this relationship between a dollar amount and the average "basket of goods" it can acquire, then the aggregate of national savings in relationship to the overall quantity of money in circulation appears fairly stable.[8] In this method, statistics scientifically certify market economies as "the spontaneous and voluntary coop-eration of individual human beings [that] need not be imposed or constructed or legislated by philosopher-kings" (1967, 92).

Friedman's theory certainly attaches itself to a mathematical frame, but so does Keynesianism. In fact, one might argue that Keynesian economics offers a more accurate mathematical picture than Friedman, who relies on the entirely imagined abstraction of "real" economic factors for which he provides no clear formula.[9] In this light, it becomes difficult to claim that the strength of Fried-man's neoliberal doctrine stems solely from its mathematical presentation. Instead, we contend that rational choice theory acquires its force from coupling the social energies housed within its core assumption—that divergent instinc-tual choices will construct a rational economic terrain as long as government does not interfere—with the exacting presentation of mathematical positivism. Fueled by his enthusiasm for Smith's "flash of genius," Friedman employed a kind of mathematical alchemy—the abstraction of reality into statistics—to establish an energetic affiliation between the invisible hand and a proper math-ematical orientation (Friedman and Friedman 1990, 13–14). Whereas attempts to gather a total economic picture and then intervene through government structures link slow and inefficient mathematical labor with an equally arduous and imprecise process of human negotiation, his precise mathematical formulas enliven the notion of irrefutable human instincts. Less a correction of Keynes's flawed proof than an invention that attends to the needs of an ever-shifting capitalist political economy, Friedman's alternative mathematical representation of the equilibrium theory directs attention to everyday economic practices. What capitalism required, and what Friedman's mathematics suggested, was that society focus its rhetorical energies on cultivating autonomous economic subjects capable of manipulating neoliberalism's increasingly sophisticated financial tools as though they were second nature. As the next section suggests, this project requires the rhetorical labor of *energeia* to actualize diverse identi-ties as economic subjects that constantly and unthinkingly adjust themselves to the fluctuations of contemporary capitalism.

The Paradox of Training Free-Market Subjects

The invisible hand of the market has enraptured the American psyche as few other metaphors have. For Rob Asen, Smith crafted "a powerfully evocative metaphor that bursts through his prose to escape the confines of political economy and obtain the status of a common sense" (2009, 1). This commonsense status requires the rhetorical labor of *energeia* in multifaceted forms. As a generative fuel, capitalist *energeia* drives ongoing disciplinary and rhetorical invention. It instantiates what Kennedy calls the "emotional energy that impels the speaker to speak" (1992, 2). Viewed from the symbolic vitality of its form, the invisible hand evokes what Kennedy describes as "the energy level coded in the message"; viewed from audience engagement, it includes "the energy experienced by the recipient in decoding the message" (1992, 2). Its rhetorical vigor both precedes (sparking economic inquiry) and exceeds (inspiring an active audience) its discursive form. For Adam Smith, the invisible hand metaphor gave semantic shape to an existing phenomenon in the world, a force previously as binding yet as unseen as "invisible *chains*" (1795, 20). The encoding of this energy into a rhetorical form "introduce[d] order into this chaos of jarring and discordant appearances, to allay this tumult of the imagination and to restore it" (1795, 20). Thereafter, whether as a flawed notion or "flash of genius," the invisible hand was to become a pressing rhetorical exigence that moved economic theorists to extensive, often impassioned engagement (Friedman and Friedman 1990, 13–14).

This evolution of the invisible hand into common sense could not be accomplished on philosophical grounds alone. The elegance of its implied equation—the precise balance between production and consumption—required economic and social analysis to be expressed through increasingly complicated forms of mathematical reasoning, which could update, quantify, and systematize the relationship between the metaphor and the actualization of its capitalist *energeia*. From the desire to capture the natural dynamics of the world by way of a universalizing rhetorical principle, to a persistent but incomplete assumption that human shortcomings can be calculated and accounted for by the government, to a descriptively hollow yet statistically valid insight that asserts the best government as no government, the mathematization of the invisible hand evolves in step with the historical ebb and flow of capitalism's self-preservation. Whereas the metaphor promotes enthusiasm for capitalism as the ultimate and effortless form of economic organization, its mathematical presentation indicates the subsequent labor required to transform this promise into a reality. Specific economic issues need to be worked out for capitalism to fulfill the promises embedded within the invisible hand, and so economists must agree on market properties such as equilibrium and optimality, the nature of self-interested

action, and the relationship of a functioning free market to government intervention. These and other queries mandate the invention, manipulation, and
revision of mathematical concepts for specific capitalist exigencies.

In order to preserve the integrity of the invisible hand (recall that self-
preservation is crucial to rhetorical energy), the nuts and bolts of bringing capitalist theory into reality require expenditures of energy beyond the disciplinary
field. Numerous public campaigns that arise out of these economic conversations reflect a further "physical energy expended in the utterance" (Kennedy
1992, 2). The distinction between the galvanizing energy that impels a speech act
(supported by disciplinary norms and warrants) and the physical expenditure
of energy that a speech act itself compels (often breaking with those norms and
warrants) is perhaps subtle. It is, nevertheless, an important distinction: while
the former directs us to apprehend the invisible hand as a *rhetorical exigence* for
economic theory, the latter orients us toward the *rhetorical effort* of promulgat-
ing the metaphor and its legitimating mathematics. The invisible hand's emphasis on the natural self-balancing mechanism of the market and the effortlessness
of the human actions that enable it demands considerable disciplinary and
political popularization in order to take root and work. Regardless of how natural the animating energies of economic neoliberalism may now feel, it did not
become the dominant model without considerable rhetorical labor.

Milton Friedman, for instance, was under no illusion that the individual
activity described by his economic platform would occur naturally. Instead, he
believed free-market subjects must be inculcated by persuasive means, a case
he makes plainly in "Neoliberalism and Its Prospects." In this short political
statement, Friedman claims that decisions emerge from an unconscious
worldview: people work *as if* they rationally weighed choices according to the
rules of that worldview, but they do so automatically and without thinking
(1951, 90). Because this automatic response is not naturally aligned with market
needs but derives from public opinion, economists need to engage public opinion before they can expect to make policy changes. Writing in 1951, Friedman
felt strongly that American public opinion, awash in Cold War anxiety and
entering an inflationary period, was poised for change. In proselytizing terms,
he declared that "it behooves us to make it clear to one and all what that [new]
faith is" (1951, 90). Taking his own advice, Friedman coauthored a series of trade
books with his wife. Among them was *Free to Choose*, which also became the
basis for a ten-part public television series that brought his anti-interventionist
arguments to ordinary Americans. Clearly, it was not sufficient to disprove

Keynesianism in the pages of economics journals. Securing the appropriate social and political conditions for the free market necessitated intensive effort, including popularization campaigns designed for maximum outreach.

Friedman's tireless public outreach and political advocacy highlight the paradox inherent in Smith's enduring metaphor. An irreducible tension exists between the purportedly instinctive behavior of capitalism and the need for pedagogical interventions that make behavior *appear* to be instinctive. Perhaps nowhere is this tension more evident in Friedman's own work than in his examples of *as if* behaviors—the leaf turning toward the sun and the billiard player executing a perfect bank shot. Both incorporate a kind of scientific precision while also capturing a sense of effortless action. However, these examples are technically quite different. A leaf moves toward the sun because it contains a light-sensing hormone (auxin) that ensures its optimal ability to photosynthesize. Human beings have no error-proof hormone for playing billiards. The billiard player must train himself to execute the perfect shot, which happens, more often than not, through trial and error, repetitive exercises, and the correction of a good teacher. Leaving this rhetorical training out of the equation, Friedman addresses only the end result: a great player and a perfect marketplace. His *as if* doctrine takes for granted that important, indeed crucial, information is exchanged through an unconscious and embodied participation in one's environment.[10] In fact, Friedman worked diligently to bring his philosophy to the public precisely because economic individuals, like billiard players, do not become habituated to new practices without repetitive instruction (Friedman 1984, 2002; Friedman and Friedman 1990).

Seeking to counter the uncertainty of actual economic practices by asserting a public pedagogy inspired by his mathematical invigoration of the invisible hand, Friedman effectively opened a pedagogical loophole. When the economy goes awry, citizens can be corralled and recalibrated to properly intuit their self-interests through the appropriate mathematical model. To put this differently, what we might call the "big E" *Energeia* of the invisible hand requires the "small e" *energeia* of its proponents, who must train economic agents to follow their natural tendencies. Friedman's mathematization of the invisible hand, as we have argued, did not reveal an objective capitalist nature as much as it directed the labor of this loophole. Regulatory responsibility lies not with governments but with individual citizens who can be expected to exercise their freedom to make instinctive decisions in market-appropriate ways. His mathematics located the individual as the appropriate site for ongoing intervention, and his public outreach demonstrates his conviction that citizens can and should be taught to behave according to natural and implicit self-calculations. Not surprisingly, there is no shortage of contemporary programs designed to teach citizens how to avail themselves of the political economic environment. The 2008 financial crisis, for instance, gave rise to an outcry for citizen-based financial literacy training. We briefly discuss this crisis not only to show that citizens must learn to acquire the particular characteristic of self-interestedness prized by neoliberal capitalism but also to emphasize such training as the rhetorical labor instigated by the most recent mathematization of the invisible hand.

Although exhortations toward a techne of financial management litter pedagogical discourses of past centuries—from mid-nineteenth-century missives on domestic economy to early twentieth-century tomes on prosperity—the campaign of the 2008 crisis evolved from the economic education movement of the 1950s and '60s. As historian Andrew Yarrow details, it was during the post–World War II era that credit and debt became facts of life and economics became the "principal language and lens through which America understood and defined itself" (2008, 397). A burgeoning fascination with economic metrics and principles found pedagogical expression in an educational platform whose rationale, "effective citizenship in an economically complex society," rested on the assumption that economic abundance was available to all (Yarrow 2008, 400). The imperative to educate citizens for an economically complex world grew more urgent in the 1980s as the economy became complicated by the sophisticated financial tools that arose as government restrictions disappeared. By acquiring greater financial literacy, we were told, individuals would seamlessly adjust their practices to this rapidly changing economic climate. Advocacy efforts coalesced around personal financial education as the training ground for economic citizenship. The well-known nonprofit Jump$tart Coalition, for instance, was founded in response to the "newly deregulated but very innovative financial system that demanded of its users a high level of financial sophistication" (Mandell and Hariharan 2004, 21A). The force of Friedman's mathematical economics (its *energeia*) exceeded the mere assertion that economic success required unencumbered capitalist practices. Instead, it encouraged the actualization of such practices. Deregulated markets enable citizens to make self-interested choices, but those citizens need to be taught how to make *good* self-interested choices.

Well before the financial meltdown, the call to train citizens as neoliberal subjects took shape in the focus on financial illiteracy. By the early 2000s, financial literacy had "gained the attention of a wide range of major banking companies, government agencies, grassroots consumer and community inter-est groups, and other organizations"—diverse stakeholders united by fear that Americans lacked the financial sophistication to choose rationally within an increasingly diversified and "abundant" financial system (Braunstein and Welch 2002, 445). This lack of financial sensibility was cited in rhetorical panics about everything from failure to participate sufficiently in the stock market to the inadequacy of retirement savings. The financial literacy movement was also entangled in promotional discourses that more generally exhorted citizens to participate widely and knowledgably as economic actors. The George W. Bush administration, for instance, promoted "financial education efforts so that fami-lies can understand what they need to do to become homeowners" ("Executive Summary" 2004). That financial literacy had been essentially conflated with homeownership as a desirable form of economic engagement was largely forgot-ten as the housing crisis unfolded and countless prominent rhetors chastised citizens for failing to educate themselves properly. Bush (2008) himself would ascribe the disaster to uninformed decision-making that allowed too many "personal financial cris[es]" to accumulate into a system-level event, rather than to the proliferation of risk in an industry increasingly unencumbered by govern-ment oversight or to his own suggestion that citizens acquire financial literacy simply by participating in the home-buying process.

The political consensus, that financial illiteracy caused the 2008 economic crisis, made financial education essential as an intervention. Financial policy-makers backed this narrative, citing the subprime collapse as crucial evidence of "how critically important it is for individuals to become financially literate at an early age" (Bernanke 2008). Established financial literacy advocates also con-firmed that the crisis had been fostered by a failure of financial knowledge, call-ing this a "teachable moment" that illustrated the importance of their cause. The crisis opened a *kairos* not for rethinking laissez-faire approaches to economic governance but for retraining the citizen to adjust to the needs of capitalism. This is in no small part the result of neoliberal mathematics, which unleashed the pent-up *energeia* of the market's invisible hand metaphor, directing labor toward financial literacy programs rather than government regulations.

Financial literacy is essentially an unbounded body of knowledge covering everything from traditional financial practices (like balancing a checkbook) to understanding financial sector innovations (like nontraditional home mort-gage products); it is an adaptable notion that can bolster the political program of economic deregulation while informing consumers how and when to act in their own self-interest. Moreover, it provides an accessible explanation when such programs fail: individuals were not appropriately instructed. The postcrisis financial literacy narrative preserved the existing political economic order against skepticism in just this way. Insisting on blame neither for markets that failed to self-correct, nor for regulatory structures to keep markets from failing, the financial literacy narrative instead makes American citizens culpable for failing to contain their "personal crises" from escalating into a national problem. In short, the financial literacy narrative provided shelter from the blowback of the Great Recession. Consonant with Kennedy's description of rhetoric's "instinct for self-preservation," ever-pervasive financial literacy calls have an energy that increases in "volume, pitch, and repetition" in relation to the impending threats against capitalism (1998, 4). The crisis not only altered the inflection and frequency of these calls, but it also positively charged such financial literacy pleas with the power to change concrete thinking and being in the world.

Financial literacy discourses employed various mathematical features to direct this neoliberal *energeia*, relying especially on the statistic (Friedman's mathematical rhetoric of choice) to identify the individual as the proper site of intervention. Although financial literacy advocates have made a habit of stockpiling persuasive statistics to advance definitions, maintain political positions, and manage public sentiment, one statistic in particular found favor during the crisis.[11] Using findings from the Jump$tart Coalition's biannual survey, critics of financially illiterate citizens repeatedly asserted that 50 percent or fewer of American adults could correctly answer a series of "basic" financial questions.[12] Readily interpreted within familiar schemas of educational success and failure— a 50 percent score is "generally considered to be a failing grade"—this statistic was widely circulated in news coverage that claimed Americans are "flunking" at finance (Palmer 2008; Ross 2011). From tapping into the fearful vision of America as an educationally substandard nation to inviting moral judgment on those who failed to uphold their mortgage obligations, the 50 percent statistic helped shift debate from lower-level stasis questions about fact and definition to value-laden concerns, leaving little room for alternative policy debates. As a resource for improving market ethos, statistics abstract complex heterogeneous behaviors into deceptively simple authoritative explanations. The illiteracy statistic, for instance, discouraged complex questions about (1) the paradigm of debt and credit underpinning notions of "positive credit behavior," (2) the accuracy of instruments that measure financial stress or adeptness, or (3) the possibility

that an inability to comprehend market futures may be a chronic part of capitalism. Not surprisingly, this widely circulating statistic inaugurated little critical
discussion about the apparent unattainability of the financially literate citizen.
We can only conclude that financial literacy aligns with neoliberalism as what
David Harvey calls a "*utopian* project"—perpetually unrealized, but no less compelling for that (2005, 19).

Indeed, the turn to financial literacy exemplifies how the crisis became an
opportunity not to reexamine neoliberalism and its economic structures but to
further instill its doctrine, including the training of natural self-interest. Perhaps indicative of its success, the financial literacy narrative exploits the same
trifecta as Friedman: resonance with the invisible hand, bolstering of mathematical arguments, and rhetorical inculcation of an injunction to recalibrate
wayward participants. None of these three contributing factors can be extracted
as a catalyst for the others because the force of capitalist *energeia* emerges from
the dynamic and historically situated configurations among all three. Clearly,
the best way to approach the relations among rhetoric, economics, and mathematics requires active and process-oriented methods rather than static and
transactional ones. We conclude, therefore, with a brief discussion of possible
shifts in our thinking about these entangled discourses.

Rhetorical Energy and the Fabrications
of Mathematical Economics

We have suggested throughout this chapter that capitalism's formative power
(its *energeia*) stems, in large part, from an historically unfolding set of associations that emerge from the mathematization of the invisible hand. Included in
this network are a series of cascading valorizations: reason over emotion, positive science over normative beliefs, individual freedom over the collective good,
and self-interest over government intervention. This assemblage contains a specific way of conceiving and enacting the relationship among its many elements.
It has a privileged mode of actualization that engages the world within what
Arjo Klamer calls "a mechanistic view of both the individual and the system"
(1987, 177). For Klamer, the combination of an embedded value system and its
formulaic deployment allows neoclassical formulations, including influential
articulations like Friedman's rational choice theory, to endure even amid set-
backs like the 2008 economic recession. Cultivated by countless pedagogical

interventions, contemporary capitalism produces purportedly spontaneous economic behaviors that do not engage deeply with the foundations of complex economic problems. It teaches individuals how to mathematically calculate economic advantage rather than how to understand the way mathematics produces and constrains economic behaviors. Fueled by capitalist *energeia* and directed by its various mathematical expressions, the invisible hand of the market promotes a kind of strict instrumentalism that delimits the possibilities of economic exploration and, consequently, economic invention.

Perhaps the constricted mathematical techne that Klamer associates with everyday practices begins in economics departments. Contemporary departments often promote a professional mathematics that replaces economic problem-solving activities with standardized mathematical knowledge. With expertise over such macroeconomic decisions as monetary production, taxation, and the distribution of resources as well as microeconomic decisions such as wage and price determinations, the professional field of economics surprisingly imagines the invention of new economic theory as superfluous to its purposes. In a 2006 study of graduate programs in economics, for instance, only 9 percent of those surveyed believed economic knowledge to be crucial to professional success. Indeed, over half those queried felt that it was unimportant. What was important, they said, was mathematics—82 percent believed that mathematical ability was very important to their future success. These beliefs are supported by a pedagogical approach focused on mathematical methods "with little attempt in the graduate programs to show how and why [particular] techniques had been developed" (Hodgson 2006, 117). Those questions that have been omitted from the pervasive authority of mathematical economics are precisely the ones necessary to enable alternative possibilities among rhetoric, mathematics, and economics. Regardless of its origins, this mechanical approach to the relationship between mathematics and economics pervades contemporary behavior and requires significant interventions.

At the very least, we need a more dynamic appreciation of how mathematical knowledge—whether contained within disciplines or seeping into public realms—constructs and constrains human behavior. Reyes underscores the importance of such rhetorical work in his study of the far-reaching consequences of Newton's and Leibniz's mathematical thinking. Their conception of calculus, he says, "demonstrates a major moment of change regarding the way humans perceived, thought about, and talked about their world" (2004, 165).

This insight ought to liberate mathematics from its imperative to unveil the world's hidden realities. In a similar way, our study of the invisible hand's mathematization seeks to emancipate mathematics from the confines of a realist style imposed on economics and other disciplines. Mathematics need not be studied as a distinct rhetorical style; nor, however, need it be studied as a self-contained field. Instead, we believe mathematical implications become most far-reaching when viewed from the lens of their entanglement with historical, social, and environmental processes, something Kennedy tried to capture in his general theory of rhetoric as an irrepressible energetic force. The relationship between rhetoric and mathematics that Reyes encourages and we support approaches mathematics as a living "tool for helping humans cope with the world" (2004, 182). Understanding mathematics as a rhetorical tool for negotiating and shaping our world opens it to greater interrogations than those fostered by the assertion that mathematical style strengthens economic credibility.

The overwhelming majority of work within the rhetoric of economics reifies both rhetoric and mathematics within forms and figures, precluding the active role of rhetorical *energeia* along with other nontraditional approaches to the persuasive nature of speech. This collection, organized around entry points into the conversation of rhetoric and mathematics, breaks out of such static conceptions of how to practice rhetorical inquiry in the disciplines. Contributing to that effort, our history of the invisible hand in economics illustrates how economics asserts authority as it borrows from mathematics, but it also suggests that the different mathematical approaches to the equilibrium theory require the invention of alternative mathematical units contributing to the consumption equals production formulation. Moreover, because this conversation remains tethered to Western capitalist traditions, its relationship between rhetoric and mathematics inflects the sociocultural and historical influences that Ed Schiappa identifies as a third avenue into the myriad intersections among rhetoric and mathematics. Economists have done more than harness mathematics to bolster realist presentations—they have invented mathematics, embedding them deeply in theories that alter not only their disciplinary terrain but also the sociohistorical and political contexts that sustain asymmetrical power relations along cultural, ethnic, and geographic lines. The rhetorical processes, such as the mathematization of the invisible hand, that give birth to particular realities and subsequently alter the shape of those realities according to continuously evolving information deserve more attention than we have so far given them.

Notes

1. Keynes felt that new mathematical contributions relied on equations to their detriment. For instance, he wrote, "too large a proportion of recent 'mathematical' economics are merely concoctions, as imprecise as the initial assumptions they rest on, which allow the author to lose sight of the complexities and interdependencies of the real world in a maze of pretentious and unhelpful symbols" (1936, 298).

2. In an interview with the *New York Times* in 1999, Heilbroner was unequivocal about what economics had sacrificed in its pursuit of mathematical validation: "The worldly philosophers thought their task was to model all the complexities of an economic system—the political, the sociological, the psychological, the moral, the historical . . . modern economists, *au contraire*, do not want so complex a vision. They favor two-dimensional models that in trying to be scientific leave out too much" (Backhouse and Bateman 2011).

3. Beed and Kane (1991) offer a more in-depth treatment of the mathematization debate.

4. Behavioral economists, following Friedman's *as if* doctrine through cognitive psychology, have been attending to these rhetorical energies since Gary Becker's work in the late 1970s. More recently, Daniel Kahneman's *Thinking, Fast and Slow* has popularized this basic rhetorical modality as "fast thinking" and says that it accounts for the vast majority of all reasoned action as it requires less effort than "slow" or carefully derived rational thinking.

5. Smith uses the "invisible hand" for the first time in "The History of Astronomy." He argues that while less advanced and less reasonable populations required the invisible hand of God to explain anomalous natural events, advanced societies call on science and philosophy to explain the ordinary course of life. He also uses the analogy in *The Theory of Moral Sentiments* and in *Wealth of Nations*. Each time Smith employs this metaphor, he references a different object, but each time the underlying idea is that an external force regulates material life so that individuals are free to pursue their purportedly natural behaviors.

6. In a recent study bemoaning the almost total absence of math-free research from economic scholarship, Daniel Sutter and Rex Pjesky (2007) pose the question "Where would Adam Smith publish today?"

7. This consumption theory can be represented as $C = A + IY$, wherein total consumption (C) results from autonomous consumption (A) and incentivized consumption (I) that depends on wages (Y).

8. In "The Methodology of Positive Economics," Friedman defines real value as "the volume of goods and services that the money will purchase" and calculates it by dividing total dollars by a price index to create a "standard basket of goods" (1953, 2). In this way, such values are both abstract—they represent how much of some hypothetical combination of goods one can purchase with a given dollar amount—and mathematically concrete in that they can be derived precisely through equations. As ever, Friedman emphasizes the empirical nature of this monetary theory by likening it to "the uniformities that form the basis of the physical sciences" (1956, 21).

9. Because Friedman's consumption function is based on a rhetorical element, permanent or lifelong income, it uses quantities that "cannot be observed directly [and] must be inferred from the behavior of consumer units" (1957, 221).

10. There exists an interesting parallelism between the automatic behavior elicited from a particular environment that Friedman discusses and the ambient rhetoric theorized by Thomas Rickert (2013).

11. The website of the California chapter of Jump$tart, for instance, presents a "compilation of facts and figures gathered in the last 24 months about America's money habits," from a range of large-scale financial literacy surveys, exhorting visitors to treat them as rhetorical

resources. "Use them," it implores, "to educate yourself, help teach others in your community, or win an argument with your Uncle Bob!" (California Jump$tart 2016).

12. The Jump$tart Survey of Financial Literacy was one of the coalition's most significant initiatives. Composed of questions deemed practical and representative of age-appropriate financial knowledge, the survey was first administered in 1997. From 2000 to 2008, it was given biannually to a stratified, random sample of thousands of high school seniors. Results found that test-takers consistently answered only around 50 percent of the questions correctly, leading to debate about the instrument and the results (Mandell 2006).

References

Aristotle. 1991. *On Rhetoric: A Theory of Civic Discourse.* 2nd ed. Translated by George Kennedy. New York: Oxford University Press.

Asen, Robert. 2009. *Invoking the Invisible Hand: Social Security and the Privatization Debates.* Michigan: Michigan State University Press.

Aune, James Arnt. 2001. *Selling the Free Market: The Rhetoric of Economic Correctness.* New York: Guilford Press.

Backhouse, Roger, and Bradley Bateman. 2011. "Wanted: Worldly Philosophers." *New York Times,* November 5. https://www.nytimes.com/2011/11/06/opinion/sunday/worldly -philosophers-wanted.html.

Beed, Clive, and Owen Kane. 1991. "What Is the Critique of the Mathematization of Economics?" *Kyklos* 44 (4): 581–612.

Bernanke, Ben. 2008. "The Importance of Financial Education and the National Jump$tart Coalition Survey." Speech 387, Board of Governors of the Federal Reserve System (U.S.). April 9. https://www.federalreserve.gov/newsevents/speech/bernanke2008 0409a.htm.

Braunstein, Sandra, and Carolyn Welch. 2002. "Financial Literacy: An Overview of Practice, Research and Policy." Federal Reserve Bulletin 88:445–57.

Bush, George W. 2008. "President Bush Announces President's Advisory Council on Financial Literacy." George W. Bush White House Archives, January 22. http://georgewbush -whitehouse.archives.gov/news/releases/2008/01/20080122-7.html.

California Jump$tart. 2019. "Financial Literacy." California Jump$tart. https://www.cajump start.org/resources/making-the-case.

Cintron, Ralph. 2010. "Democracy and Its Limitations." In *The Public Work of Rhetoric: Citizen-Scholars and Civic Engagement,* edited by John Ackerman and David J. Coogan, 98–118. Columbia: University of South Carolina Press.

Davis, Philip J., and Reuben Hersh. 1987. "Rhetoric and Mathematics." In Nelson, Megill, and McCloskey, *Rhetoric of the Human Sciences,* 53–68.

"Executive Summary." 2004. Homeownership Policy Book. http://georgewbush-whitehouse .archives.gov/infocus/homeownership/homeownership-policy-book-execsum.html.

Friedman, Milton. 1951. "Neo-Liberalism and Its Prospects." *Farmand* 17:89–93.

———. 1953. "The Methodology of Positive Economics." In *Essays in Positive Economics,* 3–43. Chicago: University of Chicago Press.

———. 1956. "The Quantity Theory of Money—A Restatement." In *Studies in the Quantity Theory of Money,* edited by Milton Friedman, 3–21. Chicago: University of Chicago Press.

———. 1957. *A Theory of the Consumption Function.* Princeton: Princeton University Press.

———. 1967 "Value Judgments in Economics." In *Human Values and Economic Policy: A Symposium*, edited by Sidney Hook, 85–93. New York: University Press.

———. 1984. *Tyranny of the Status Quo*. Boston: Houghton Mifflin.

———. 2002. *Capitalism and Freedom*. Chicago: University of Chicago Press.

Friedman, Rose, and Milton Friedman. 1990. *Free to Choose*. New York: Harcourt.

Harvey, David. 2005. *A Brief History of Neoliberalism*. Oxford: Oxford University Press.

Heilbroner, Robert L. 1953. *The Worldly Philosophers*. New York: Touchstone.

Hodgson, Geoffrey M. 2006. *Economics in the Shadows of Darwin and Marx: Essays on Institutional and Evolutionary Themes*. Northampton: Edward Elgar.

Huntington, Samuel. 1968. *Political Order in Changing Societies*. New Haven: Yale University Press.

Kahneman, Daniel. 2011. *Thinking, Fast and Slow*. New York: Farrar, Straus and Giroux.

Kennedy, George A. 1992. "A Hoot in the Dark: The Evolution of a General Rhetoric." *Philosophy and Rhetoric* 25, no. 1:1–21. https://www.jstor.org/stable/40238276.

———. 1998. *Comparative Rhetoric: An Historical and Cross-Cultural Introduction*. New York: Oxford University Press.

Keynes, John Maynard. 1921. *A Treatise on Probability*. London: Macmillan.

———. 1936. *The General Theory of Employment, Interest, and Money*. New York: Harcourt, Brace.

Klamer, Arjo. 1987. "As If Economists and Their Subject Were Rational." In Nelson, Megill, and McCloskey, *Rhetoric of the Human Sciences*, 163–83.

Klamer, Arjo, and Deirdre McCloskey. 1998. "Economics in the Human Conversation." In *The Consequences of Economic Rhetoric*, edited by Arjo Klamer, Deirdre N. McCloskey, and Robert M. Solow, 3–20. Cambridge: Cambridge University Press.

"Making the Case for Financial Literacy 2016." 2016. California Jump$tart. https://www.cajumpstart.org/resources/making-the-case.

Mandell, Lewis. 2006. "Financial Literacy: If It's So Important, Why Isn't It Improving?" *Networks Financial Institute Policy Brief No. 2006-PB-08*. http://dx.doi.org/10.2139/ssrn.923557.

Mandell, Lewis, and Maykala Hariharan. 2004. "A Dream of Financial Literacy." *Credit Union Magazine*, January, 21–22A.

McCloskey, Deirdre N. 1998. *The Rhetoric of Economics*. Madison: University of Wisconsin Press.

Nelson, John S., Allan Megill, and Donald N. McCloskey, eds. *The Rhetoric of the Human Sciences: Language and Argument in Scholarship and Public Affairs*. Madison: University of Wisconsin Press.

Palmer, Kimberly. 2008. "The Financial Literacy Crisis." *U.S. News and World Report*, April 2. https://money.usnews.com/money/personal-finance/articles/2008/04/02/financial-literacy-101.

Reyes, G. Mitchell. 2004. "The Rhetoric in Mathematics: Newton, Leibniz, the Calculus, and the Rhetorical Force of the Infinitesimal." *Quarterly Journal of Speech* 90:163–88. https://doi.org/10.1080/0033563042000227427.

Rickert, Thomas. 2013. *Ambient Rhetoric: The Attunements of Rhetorical Being*. Pittsburgh: University of Pittsburgh Press.

Ross, Janell. 2011. "Americans Struggle with Financial Problems They Don't Understand." *Huffington Post*, June 9. https://www.huffingtonpost.com/2011/06/09/financial-ignorance-and-denial_n_873181.html.

Smith, Adam. 1795. "The History of Astronomy." In *Essays in Philosophical Subjects*, edited by
    Joseph Black and James Hutton, 1–93. London: Cadell and Davies. https://archive
    .org/details/essaysonphilosooostewgoog/page/n8.
———. 1991. *Wealth of Nations*. Amherst: Prometheus Books.
———. 2009. *The Theory of Moral Sentiments*. Abingdon: Glass House Books.
Sutter, Daniel, and Rex Pjesky. 2007. "Where Would Adam Smith Publish Today? The Near
    Absence of Math-Free Research in Top Journals." *Econ Journal Watch* 4, no. 2:230–40.
Westin, Monica. 2017. "Aristotle's Rhetorical Energeia: An Extended Note." *Advances in the
    History of Rhetoric* 20 (3): 252–61. https://doi.org/10.1080/15362426.2017.1384769.
Wynn, James. 2009. "Arithmetic of the Species: Darwin and the Role of Mathematics in His
    Argumentation." *Rhetorica* 27 (1): 76–97. https://www.jstor.org/stable/10.1525/rh
    .2009.27.1.76.

4

The Horizons of Judgment in Mathematical Discourse |
Copulas, Economics, and Subprime Mortgages

G. Mitchell Reyes

Capitalism and mathematics are intimately related: mathematics functions as the grammar of techno-scientific discourse that every form of capitalism has relied on and initiated. So it would be feasible ... to see in the realist account of mathematics an ideological formation serving certain (techno-scientific) ends within twentieth-century capitalism.
—Brian Rotman

Rhetorical scholars have long studied economic rhetoric. In the 1980s Deidre McCloskey's groundbreaking work gave rise to a productive and insightful subfield of rhetorical studies.[1] Much of this work is carefully argued, illuminating the importance of democratic deliberation within economic rhetorics, the influence of persuasion and value judgments in economic decision making, the ways economic processes exceed the boundaries of the rational-actor model, and the play of mimesis and circulation in the cascading activations that characterize so many economic processes. Given rhetoric's historical emphasis on speech, argument, and persuasion, it is not surprising that these studies tend to focus on traditional discursive practices within economics, that is, on speeches given by prominent actors (such as Alan Greenspan) or on argumentation within economics and among economists (policy debates, for instance) or on the circulation of certain influential ideographs as carriers of conflicting ideologies.[2] As with all tendencies, one can find exceptions to the rule (McCloskey's work first among them), yet the preponderance of research on economic rhetoric focuses on conventional rhetorical forms of economic discourse.

This essay builds upon and expands studies of economic rhetoric with attention to "the numeric," which names a whole field of rhetorical action within economics that often supports the traditional rhetorical forms (speeches, written arguments, debates) more frequently featured in our journals. Most modern economic discourses, for instance, deploy modalities of numeracy (statistics, econometrics, mathematics) in order to buttress and reinforce their economic arguments.[3] A grounding assumption for this essay is that if rhetorical scholars hope to significantly influence discourse and practice within economics, we must learn to engage critically with the mathematical discourses that form the foundations of most modern economic arguments.

To begin such an ambitious endeavor (one far beyond the powers of a single academic article), I suggest we turn to two allied resources, one from within rhetoric and one from outside of it: the first emerges from a small but growing body of research into rhetoric and mathematics, a topic that emerged from the same now famous 1984 Iowa symposium that birthed the rhetoric of economics (see Nelson, Megill, and McCloskey 1987; Davis and Hersh 1987). Scholars in this collective are increasingly finding connections between rhetoric and math that challenge the old modernist opposition between the two (see Reyes 2014; Wynn 2012; Rotman 2000). From this research, we can borrow ideas and methods for reading the rhetoric *in* mathematical practice, and, ultimately, we can learn how to reassemble the horizons of judgment that inform mathematical propositions. As we will see, those horizons of judgment are well hidden (which is why I use the metaphor *horizon*—something always present but seemingly out of reach), yet their content is crucial for understanding and critically engaging with the myriad economic, social, and cultural consequences of mathematical discourse. The second resource emerges from the collected writings of Bruno Latour, who offers an innovative way to think about and study mathematical discourse as a networking modality that materially expands our social collectives. Combining these two resources gives rise to a hybrid analytic equal to the task of engaging with the numerical discourses that form the bases of modern economic arguments even as it challenges rhetorical scholars to think differently about certain disciplinary concepts such as rhetoric, agency, and power.

To explore the numerical discourses of modern economics and how they support economic claims, I examine their role in the 2008 financial crisis.[4] I look specifically at the rise of a particular mathematical formula known as the Li Gaussian copula, which played a significant role in the spread of subprime mortgages leading up to the 2008 collapse. Unraveling this copula reveals the constitutive rhetorical force of mathematical discourse—its capacity to invent, accelerate, and concentrate economic networks—but also the fact that inside every mathematical statement lies a horizon of judgment that, when used, effectively displaces practical judgment with the judgments already baked into the mathematical statements employed.[5] Those horizons of judgment, however, are often concealed (both intentionally and unintentionally) within forms of mathematical realism that transform mathematical models with domains of validity (and invalidity) into hard *technologies* for economic production and extraction. This essay thus not only examines the constitutive forces of mathematical discourse within a particular case but also demonstrates how rhetorical analysis can effectively reassemble the horizons of judgment so well hidden within technical economic discourse.[6] Doing so will, I hope, put rhetorical scholarship in a better position to offer not merely a critique of economic positivism but, more importantly, a productive alternative to it.

The term "copula" derives from the Latin word *copulare*, to connect or to join. By 2008 David X. Li's Gaussian copula had connected a great variety of financial entities, including most of the major investment firms, nearly all large banks, and even several key government rating agencies. He had, in the words of Felix Salmon, cracked "a notoriously tough nut—determining correlation, or how seemingly disparate events are related . . . with a simple and elegant mathematical formula" (Salmon 2009, 16). That formula, originally published in Li's "On Default Correlation" in 2000, quickly vaulted Li into the economic limelight. He became *the* master "Quant"—a name used for financial engineers since the 1980s. He moved from the Canadian Imperial Bank of Commerce in Canada to a subsidiary of J.P. Morgan in New York to director of derivative investing at Citigroup, all in a few years. Many expected him to receive the Nobel Prize in Economics for his work.

At a national financial engineering conference in 2003, he was regarded with esteem: "In front of a room of hundreds of fellow Quants . . . he ran through his model—the Gaussian copula function for default." "The presentation," Sam Jones reports, "was a riot of equations, mathematical lemmas, arching curves and matrices of numbers. The questions afterwards were deferential, technical" (Jones 2009). Li's copula was considered an ingenious breakthrough, a means of finally predicting with mathematical rigor the default risk between seemingly disconnected debt obligations (i.e., loans). His formula was used not only to determine default risk but also to price that risk so that it could be sold (as credit default obligations [CDOs]) and to rate the risk of investment of bundled debt obligations. Alan Greenspan, convinced of the mathematical veracity of the results of Li and other top Wall Street Quants, claimed in 2005, "recent regulatory reform, coupled with innovative technologies, has stimulated the development of financial products . . . that facilitate dispersion of risk. These increasingly complex financial instruments have contributed to the development of a far more flexible, efficient, and hence resilient financial system" (Greenspan, in Krugman 2013, 54).

The vast majority of financial experts saw Li's formula as an effective means of predicting, pricing, and rating risk. It was quickly integrated into the global financial system, for both *determining and rating* investments. This integration, however, led to the spread of subprime mortgages and the eventual implosion of the global economy. How did this come to pass? Why did the global financial industry embrace Li's formula so quickly? How did that formula—a Gaussian *model* for *forecasting* default risk—become reified into an *"innovative technology"* of risk *measurement?*[7]

To address these questions, I examine Li's original article, combining modes of analysis of mathematical discourse from rhetorical studies (constitutive theory, close textual analysis, and analogical deconstruction) with Bruno Latour's actor-network theory. Together, these reading strategies form a broader analytic that attends to both the rhetorical modalities of mathematical discourse (its powers of assemblage, concentration, and scalability) and the argumentative structures within which those modalities are often concealed. Close examination of Li's "On Default Correlation" reveals how mathematical discourse in combination with a positivist/realist style shaped Li's argument, eventually concealing the central analogy and formative assumptions that constituted the copula's horizon of judgment. One consequence of this work was the displacement of practical judgment with a form of technological rationality that concentrated the networks of global finance around a single node (the copula), *rendering those networks both larger and, contra Greenspan, more fragile.*

Developing these arguments calls first for construction of the rhetorical-Latourian hybrid analytic. Only then will we be in position to reassemble the horizon of judgment within Li's copula, see the artistry therein, and examine the vast financial networks that emerged and eventually collapsed as a result of the copula's spread.

Hybrid Analytic: Rhetoric, Mathematics, and Latour

How can one think of math rhetorically? Is math not the antithesis of rhetoric, the former dedicated to unassailable truth and the latter to modes of persuasion? Isn't math the great translation machine that carries us from the subjective quicksand of politics and ideology to the solid epistemic shores of objective knowledge? These questions were in fact the very ones that Philip Davis and Reuben Hersh posed in their pioneering work on rhetoric and mathematics in 1984. Ahead of their time, Davis and Hersh dared defy academic orthodoxy, arguing that sometimes math is rhetorical, just as on occasion rhetoric is mathematical (1987, 54). When they said "math is rhetorical," they meant that mathematics can be used to obfuscate and/or create the appearance of objectivity when none exists, and in this sense rhetoric functioned for them as a synonym for manipulation. When they said "rhetoric is mathematical," on the other hand, they emphasized how ideas are communicated within the classroom as well as between experts, and in this sense rhetoric took on the less pernicious (even potentially productive) sense of persuasion within an intellectual community. Theirs was a conventional neo-Aristotelian notion of rhetoric (Reyes 2004, 166), to be sure, yet Davis and Hersh's work nevertheless opened the door to rethinking the relations between rhetoric, mathematics, and their potential entanglements.

Rhetorical scholars have come some distance since Davis and Hersh's first provocations, examining everything from the consequences of mathematization in biology and public health (Wynn 2012; Mudry 2009) to the ways quantification helped Manhattan Project scientists dissociate from the ethical quandaries of their work (Jack 2009) to the role rhetoric played in the emergence of the calculus (Reyes 2004). The themes set out by Davis and Hersh certainly find their place within this scholarship, it being naturally concerned with how mathematization functions as an appeal to authority or enables the selective manipulation of data, yet this work also goes well beyond what Davis and Hersh originally had in mind.

Perhaps the most significant advance entails a theoretical shift in the understanding of rhetoric itself. Whereas for Davis and Hersh (both trained mathematicians) rhetoric is limited to the domains of intentional persuasion and/or manipulation (and thus fits squarely within the neo-Aristotelian paradigm), for contemporary scholars rhetoric is *constitutive*. The difference here is hard to overstate: if rhetoric remains the province of intentional persuasion, one might reasonably ask how appeals to logos, ethos, or pathos happen within mathematical contexts without ever implicating mathematical concepts themselves, which remain reassuringly arhetorical. As I have argued in previous articles, within the early scholarship on rhetoric and mathematics "the argumentative logics that constitute mathematics' productive vocabulary (numbers, geometric figures, axioms, the Cartesian coordinate system, etc.) remain unquestioned while the argumentative strategies between mathematicians and the rhetorical influence of mathematics in other discursive fields take center stage" (Reyes 2004, 166–67). It is only the communication of mathematical concepts, in other words, and not their invention, that would be of proper interest to the neo-Aristotelian critic.

The shift to rhetoric as constitutive transforms the scene dramatically. Rhetoric becomes not just a vehicle of representation and communication but the very material out of which new mathematical concepts emerge. Rhetoric here reflects the many advances of twentieth- and twenty-first-century theory, rendered not just as intentional argument but as symbolic action in toto, which includes radical theories of rhetoric not simply as means for symbolic constructions of reality (à la social constructivism) but rather as the space where many realities are fashioned, a space where language and symbols themselves (often far in excess of the intentions of the humans that mobilize them) have agency. And it is those agencies that in part help constitute the interference patterns that emerge between conceptual and material realms, giving rise to the realities in which we are enmeshed.[8]

Constitutive rhetoric is a material-conceptual orientation from which to see mathematics (as well as other discursive formations) anew. Mathematical objects are typically conceived as existing independent of human cognition. Mathematical symbols are, accordingly, more or less adequate representations of ideal mathematical objects, and the purpose of doing mathematics, as Brian Rotman—a mathematician who studies the semiotics of math—suggests, is to discover "objective irrefutably-the-case descriptions of some timeless, spaceless, subjectless realm of abstract 'objects'" (2000, 30). From this perspective, which is broadly known as "mathematical realism," mathematical discourse is always secondary to genuine mathematical thought and the ideal objects that thought contemplates. As such, any role that rhetoric might have regarding mathematics is largely parasitic, at best helping spread mathematical truths and at worst actively obscuring them. This is because, for the mathematical realist, symbolic action is merely an imperfect medium for the representation of a priori mathematical objects. As Rotman concisely summarizes, "language, for the realist, arises and operates as a name for the preexisting world. Such a view issues in a bifurcation of linguistic activity into a primary act of reference—concerning what is 'real,' given 'out there' within the prior world waiting to be labeled and denoted—and a subsidiary act of describing, commenting on, and communicating about the objects named" (2000, 30). A shift to rhetoric as constitutive, however, challenges this realist paradigm and the metaphysics that traffic with it. Instead of considering mathematical statements as better or worse representations of preexisting mathematical objects, a constitutive approach encourages scholars to imagine symbols, inscriptions, and arguments as the material out of which mathematical concepts emerge, coalesce, and are (if successfully articulated) integrated into the existing mathematical lore. Thus rhetoric plays a significant role not just in the communication, manipulation, and/or dissemination of math but in the very practices of innovation and invention within mathematics. Rhetoric, as embodied material practices of inscription, is not an enemy, an obstacle, nor a mere vehicle of mathematics but rather an engine of its evolution. From a constitutive perspective, what is most interesting about math is not how it reveals the truths of an a priori reality but how mathematical propositions and apparatuses (such as the golden ratio or the Cartesian coordinate system) actively constitute realities as they emerge and interact with our social-material worlds.[9]

However promising the constitutive approach to math, questions persist. If, as the constitutive approach claims, rhetoric is the material out of which new mathematical objects emerge, how exactly does that happen? Surely argument alone is a necessary but insufficient condition. Equally vexing, if those conceptual objects really do reconstitute realities as they interact with those realities, how do we explain such alchemy? How, in short, does the language of math become materially manifest in the world? Bruno Latour's work might seem a strange place to find answers to these questions, preoccupied as he is by science studies, by the construction of his actor-network theory (ANT), and by attention to subjects such as biology and chemistry far more than the mathematical sciences. When gathered together, however, Latour's engagements with math (which are spread throughout his many projects) offer several productive ways to *trace the materiality of mathematical discourse and its many material consequences.*[10]

Following Latour's lead, let us begin with a vignette: King Hiero sits on his throne—troubled by the forces threatening Syracuse, his people, his empire—when a letter is delivered from a young man named Archimedes, who (according to Plutarch) made the astonishing claim that "with any given force it was possible to move any given weight" (Plutarch, *Life of Marcellus* 14.7).[11] The young Archimedes was so bold, tells Plutarch, that he professed, "if there were another world, and he could go to it, he could move this" (14.7). Astonished by the claim, King Hiero requested a demonstration, leading to the oft-told story of Archimedes raising a "three-masted merchantman of the royal fleet" without any great effort by "setting in motion with his hand a system of compound pulleys" (14.8). Struck by the potential of his art, King Hiero immediately set Archimedes to work at designing "offensive and defensive engines to be used in every kind of siege warfare" (14.9).

However incomplete Plutarch's account, for Latour it reveals several interesting things about mathematics. Math, according to convention, is the language of abstract thought, the contemplation of the common forms that lie beneath the appearance of things; it is that which transports us from the realms of belief and opinion to the realms of truth and necessity. As Bertrand Russell once opined, "mathematics takes us still further from what is human, into the region of absolute necessity, to which not only the actual world, but every possible world, must conform" (1907, 30–42). Russell's mathematical realism, which is the dominant narrative of math since Plato, tells of a world of abstract but absolute and unchanging objects hidden from quotidian view. Math here becomes an escape from and a corrective to the whims and follies of politics, the decisions of which are based on mere opinion (Nussbaum 2001, 110). But from the contours of Plutarch's account (and the many other historical accounts like it) Latour assembles a different mathematics, one that does not escape but rather *extends* politics; a practice of thinking that assembles powerful apparatuses of translation through which our social and political collectives *materially expand*. From a Latourian perspective, Archimedes did not so much reveal to King Hiero the secret power of ratios exercised through the technology of compound pulleys; instead, he massively transformed "power relations through the intermediary of the compound pulley"; in doing so, "he also reversed political relations by offering the king a real mechanism for making one man physically stronger than a multitude" (Latour 1993, 110). Before the compound pulley, the Sovereign, while representative of his people, was not stronger than his people. After the compound pulley, the Sovereign—allied with a new technology—was suddenly more formidable, and thus perhaps less indebted to his subjects for his power. How to make sense of this moment of *empowerment*? One could, as Plutarch himself does, tell a story of transcendence, Archimedes becoming the sage, tapping the secrets of nature written in mathematical code. Doing so gives rise to a clear hierarchy between math and politics: the latter indebted to the former, the former purified of the latter. Latour offers us a different reading.

What is fantastic about mathematics for Latour is how it accomplishes the opposite of what we are so often told math does. Instead of allowing humans to transcend the terrestrial realms of the political and the social, math in fact enables the extension of the political and the social through novel alliances. In the Hiero-Archimedes story, we find an *emergent alliance between a political form and the compound pulley that materially transforms the social collective*. King Hiero's power has expanded not merely through Archimedes's genius but also through a new association between humans and nonhumans (via the compound pulleys and reengineered siege engines) that Archimedes's mathematical propositions of ratios made possible. But for Latour those mathematical propositions do not reveal an a priori law of nature, and Archimedes did not "discover" said law. Thinking in these traditional metaphysical ways only apotheosizes math and mathematicians while concealing from view the practice of mathematics as a practice of assemblage, one that far from separating humans from nonhumans or society from nature in fact breeds hybrids of humans and nonhumans, societies and natures, materially expanding our social collectives in the process.

Everything sensible seems simple once said, but the difference between this Latourian understanding and the conventional realist understanding of math is profound. Realists, as we know, make the ontological presumption that mathematical objects exist prior to any human contemplation of them. Ontologically, they are absolute beings that transcend all historical and environmental change; they are, as Russell opined, that "to which not only the actual world, but every possible world, must conform." Change is thus an illusion, an appearance that conceals the unchanging truths that lie beneath, which only humans (that highest of being) can discover through the forms of pure reason that gave rise to math in the first place.[12] In contrast, a Latourian approach to math rejects (at least initially) all ontological presumptions and instead begins with practice, seeking to understand not what mathematics "is" but *what mathematics does, how it works, how those who think mathematically practice their art*. In doing so he finds that math is not that to which all things (human, nature, or machine) must conform but rather a practice of translation that renders what was once incommensurable commensurable. In this light, Archimedes's propositions were powerful because they rendered commensurable exactly what was, prior to the compound pulley, incommensurable: politics and math, the one and the many. The consequences were substantial: "Up to that time, the Sovereign represented the masses whose spokesperson he was, but he had no greater strength as a result. Archimedes procured a different principle of composition for the

Leviathan by transforming the relation of political representation into a relation of mechanical proportion. Without geometry and statics, the Sovereign had to reckon with social forces that infinitely overpowered him. But if you add the lever of technology to the play of political representation alone, then you can become stronger than the multitude" (Latour 1993, 110). The two key phrases here are "principle of composition" and "transforming the relation," for they capture two points essential to understanding *how* math works for Latour. Point one: *mathematical propositions are principles of composition*; they are not representations of transcendental truths. They are actors that enact a recomposition of the existing collective. This means that mathematical propositions have an agency unto themselves that is excessive to human agency.[13] Archimedes was certainly aware of how his mathematics of ratios made commensurable the incommensurability of the large and the small, but one would be hard pressed to claim he foresaw how those same propositions would recompose the relations of power between sovereign and citizen. These unintended reverberations—or, following Karen Barad, "diffractions" (2007, 71–94)—are traces of the agency of mathematical propositions. Understanding mathematical propositions as actors (or actants) renders any claim that they reflect an a priori reality nonsensical, since they so clearly transform and extend the collective of humans and nonhumans that we call reality. In order to understand how math works as a networking modality, then, Latour suggests we must forgo the metaphysical logics of representation for the modalities of translation and mediation.[14]

Translation and mediation bring us to point two: like all discursive formations that perdure, mathematical discourse is a powerful system of translation and mediation out of which new hybrids emerge, and those hybrids can, under the right circumstances, "transform the relations" of the networks that compose our world. If we want to understand how mathematical discourse becomes materially manifest, we cannot begin with the presumption of the a priori object, which merely conceals from view the material practices of inscription, translation, and assemblage that constitute mathematics. Instead, "we start from the *vinculum* itself, from passages and relations, not accepting as a starting point any being that does not emerge from this relation that is at once collective, real, and discursive" (Latour 1993, 129). What careful rhetorical study of mathematical practice teaches us (perhaps better than any other discursive form) is that there are not two worlds, one made of symbols and one made of things, but one world of relations, and that, while some of those relations certainly existed prior to any human that might contemplate them (the relations between oxygen and hydrogen that we call "water," for instance), many others have emerged through the practices of inscription and symbolic action that bind humans and nonhumans together in increasingly novel ways.

How exactly do these practices of inscription give rise to novel relations? Latour offers several examples. Consider, he suggests, how numbers symbolically translate many diverse phenomena into a single commensurable *form*: "*Numbers* are one of the many ways to sum up, to summarize, to totalize—as the name 'total' indicates—to bring together elements which are, nevertheless, not there. The phrase '1,456,239 babies' is no more made of crying babies than the word 'dog' is a barking dog. Nevertheless, once tallied in the census, the phrase establishes *some* relations between the demographers' office and the crying babies of the land" (1987, 234). The demographers' office, one center of calculation among many, uses numbers to "know" a population. But the power of mathematizing a population does not lie in the numbers themselves; it lies in the concentration of diverse phenomena—age, gender, wealth, religious affiliation—into *one form*: "Were we to follow how the instruments in the laboratories write down the Great Book of Nature in geometrical and mathematical forms," Latour declares, "we might be able to understand why forms take so much precedence. In centers of calculation, you obtain paper forms from totally unrelated realms but with the same shape (the same Cartesian coordinates and the same functions, for instance). This means that *transversal* connections are going to be established in addition to all the *vertical* associations made by the cascade of rewriting" (1987, 244). Numbers, then, are one of the first technologies of mathematization, which is a practice of *vertical and transversal rewriting*, of translating the world into mathematics—into a formal language that renders commensurable what once appeared incommensurable. Out of that process many new potential relations emerge. They are *potential relations* precisely because they *do not yet exist*. But once the census data are collected and analyzed one can begin to link the number of babies in the land with something like fertility rates, which might correlate with pollution or the size of public parks or the quality of schools. Numbers, in short, *manufacture* a commensurability of form that encourages the humans that interact with them to *imagine* novel relations, unconstrained as they are by the radical heterogeneity of everyday life. Those humans can then take those novel relations to those in power (just as Archimedes did), and those novel relations can then lead to the creation of new hybrids, new machines, new institutions that ultimately reconfigure the social/material world.

European exploration of the Pacific in the eighteenth century offers one last example to underscore how mathematical discourse materializes in the world. When Lapérouse visited Segalien (or Sakhalin) in the Pacific Ocean in 1787 he was weak: he had no knowledge of the land, the navigable straits, or the points of danger; he was dependent on the native population for guidance. Yet when Europeans returned a decade later they were stronger—no longer dependent on those same natives. What changed in ten years? The modalities of number and calculation, combined with the Cartesian coordinate system, allowed explorers to extract and mobilize traces through the use of logbooks; those traces, slowly and painstakingly gathered in faraway centers of calculation, enabled the production of navigational maps; those maps facilitated flows of capital, extraction of resources, and exploitation of peoples. Through numbers and Cartesian coordinates and logbooks Segalien was subsumed (along with many others) into a broad category of European cartography—in French, *Pacifique oriental*; in English, the East or West Pacific—none of which named places as much as they did new relations of power we now call colonialism. And colonialism is in part a name for the desire to control at a distance: "How to act at a distance on unfamiliar events, places and people? Answer: by *somehow* bringing home these events, places, and people" (Latour 1987, 223). Numbers, Cartesian coordinates, and logbooks combined to form an apparatus that transformed the ragged unknown coastlines of Segalien into the stable forms of East and West Pacific navigational maps. How? The inscriptions in the logbooks, through number and an agreed upon system of coordinates, became "immutable mobiles," which traveled back from Segalien to the centers of calculation in Europe, which then allowed European scientists to create simulacra (navigational maps) of what they increasingly referred to as "the East [or West] Pacific" (242–47). Those simulacra have a number of material consequences: (1) they allowed Europeans to *simulate* their next exploration before ever leaving shore; as a result, (2) those Europeans became less indebted to the peoples of Segalien (and elsewhere) for their safe passage, (3) the heterogeneity of the peoples and places inhabiting these vast geographies become transformed into the stable forms of the simulacra, (4) the simulacra, in tandem with other cultural forces, slowly translate the subjectivities of myriad peoples into the objectivities of East and West Pacific navigational maps, and (5) those objectivities could then be more easily *located, extracted, and sold.*

The point of these examples for Latour is not to condemn math for reductionism, which would be a massive misunderstanding of how it works, but on the contrary to try to understand the unparalleled productivity of the mathematical sciences.[15] "The sciences multiply new definitions of humans without managing to displace the former ones, reduce them to any homogenous one, or unify them. They add reality; they do not subtract it" (Latour 1993, 137). Just like Archimedes's compound pulleys, European navigational maps were in the eighteenth century novel hybrids that materially expanded the collective of humans and nonhumans that existed at the time.

Thinking of mathematical discourse in this fashion advances the rhetorical study of math in several productive ways. Through a neo-Aristotelian approach scholars study how mathematicians intentionally persuade one another as well as how math is strategically used (by both mathematicians and nonmathematicians) to obfuscate or manipulate a rhetorical situation. Through a constitutive approach scholars examine the powers of mathematical subjectification and the role symbolic action and argumentation play in the invention of new mathematical concepts. But with a Latourian approach we can extend beyond the realms of the symbolic and the conceptual and study math as a practice of conceptual/discursive/material assemblage. The task of a rhetorical study of mathematics shifts accordingly, focusing not on suasion in or through math but on how mathematical discourse translates the existing relations of a collective into new relations, and how those new relations form novel hybrids of humans and nonhumans and ultimately illuminate the consequences of those relations for the collectives we find ourselves in.[16]

None of these approaches in isolation, however, would be up to the task of examining the horizons of judgment within technical economic discourse. Those horizons are diverse phenomena, full of intention, motive, analogy, delimiting assumptions, human/nonhuman hybrids, vertical and transversal relations, commensurabilities, and domains of validity; and that does not begin to mention how well these horizons are hidden from view, buried through strategies of argument and various appeals to realism (some subtle, some overt). This again is why I use the metaphor of *horizon*—something always far off, present but intangible. Rather than take a strictly Latourian approach, then, we need to draw from all of our resources—from a neo-Aristotelian attention to argument structure and appeal, from a constitutive interest in symbolic action as the substance of mathematical invention, and from a Latourian focus on practices of translation that actively transform problem-situations and create space for new hybrids to emerge. Only through this imbricated reading strategy can we begin to bring the horizon of judgment within Li's copula into view, see how it displaced practical judgment, trace the ways it translated and extended existing economic networks, and thereby understand it as an apparatus (like the compound pulley or navigational maps) that mediated the recomposition of global finance.[17]

The Li Copula: Principle of Composition

Within six years of publication, the Li copula had significantly transformed financial markets, and it was often described by market participants as *"beautiful, simple,* and, most commonly, *tractable."* Stanford University finance professor Darrell Duffie noted that by 2006 "'the corporate CDO world relied almost exclusively on this copula-based correlation model'" (Salmon 2009, 19). And that "corporate CDO world" was growing exponentially—from 275 billion in 2000 to 4.7 trillion in 2006. Why did Li's copula have such a dramatic impact? What problem did it solve that eventually unleashed these massive flows of capital?

To begin to address these questions, we must first understand the problem-situation that Li and the rest of quantitative finance faced at the end of the twentieth century. That story begins with a practice called "tranching," which arose in the 1970s as a way for large banks to create additional revenue streams. In short, tranching (a technique still practiced) allows a bank that holds a collection of mortgages to gather them into a mortgage pool and then sell them as mortgage-backed securities (MBSS) to interested investors.[18] These MBSS are often tranched, or "sliced" into different categories of risk. Those categories typically range from AAA to unrated, with many tranches (AA, A, BBB, BB, etc.) in between. AAA tranches are the least risky and offer the smallest return on investment. Unrated tranches have the greatest risk but also the greatest return.[19] Tranching enables banks to sell their mortgages but continue servicing them— thus continuing to profit while reducing risk exposure.[20] However, with tranching comes the problem of pricing risk: imagine you are an investor considering purchase of an MBS. What is a fair return on investment for a AAA, BBB, or an unrated MBS? For several decades this was a significant problem within quantitative finance. Prior to Li's copula the techniques for pricing risk were considered approximate at best, in large part because no one had found a way to accurately predict default correlation between loans within a tranched MBS (what Li would later call a "credit derivative").

Why was default correlation such a difficult problem for the best minds in quantitative finance? Consider this simplified scenario: You own a house in Portland, Oregon. If you default on the mortgage, will that increase the chances of your neighbor defaulting on her mortgage? If so, by how much? What, in other words, is the default correlation between these two mortgages? Even more difficult, what is the default correlation between your mortgage and another random mortgage in Pennsylvania, New York, Alabama, or any other state for that matter? The answer is, *it depends*—it depends on the specifics of the relationships, or lack thereof, between homeowners, local and national housing markets, and national and global economies at a particular historical moment. Even this simple hypothetical, then, reveals the highly contextual nature of default correlation, which in mathematical terms means that it is both highly unstable (possessing high degrees of variance) and extremely intractable (unable to be easily controlled and therefore unable to be modeled with much accuracy or reliability).[21]

The intractability of default correlation was a major problem within structured finance. Throughout the 1980s and 1990s Wall Street Quants deployed a number of strategies to approximate default correlation between mortgages, including analysis of historical default data coupled with practical diversification strategies, but these were admittedly crude. As a result, rating agencies such as Moody's and Fitch had an understandably conservative posture toward MBSS, granting high ratings to only the safest "senior" tranches. This in turn constrained the size of the market for MBSS, since the major players Fanny Mae and Freddy Mac—as government-sponsored organizations—could only purchase "investment grade" securities (those with a BBB rating or higher). Within this environment, banks could pool their mortgages and create tranched MBSS, but they could only sell the highest-rated tranches to major market investors, and because of the uncertainty of the default correlation, only a small portion of those tranches were given high ratings. The remainder of the MBSS then had to be sold by way of what Coval, Jurek, and Stafford call "'private-label' mortgage-backed securities" (2009, 16)—a much smaller market with less favorable terms for the banks.

The problem with default correlation, in the end, did not have to do with risk but with uncertainty. Investors do not mind risk, assuming they can reliably assess and price that risk; what they dislike is uncertainty. And default correlation was the poster child for uncertainty before the Li copula. Prior to that copula's emergence, circulation, assimilation, and eventual translation of global finance, investors used the admittedly crude tools of practical judgment to guide investment decisions. They examined the actual mortgage assets that backed the tranched MBSs, and they considered the relative diversity of those mortgages as a heuristic for judging their degree of independence and thus insulation from or exposure to default correlation, but these methods of judgment were themselves intractable—hard to control, human dependent, and time-consuming to implement. The Li copula, however, rendered these forms of practical judgment (and their inefficiencies) superfluous, offering in their place what appeared to be a mathematically rigorous way to determine default correlation.[22]

The Analogy at the Heart of the Copula

How did the Li copula transform something as intractable and unstable as default correlation into what Latour would call an "immutable mobile," an object perceived to be "easily at hand and combinable at will," something that could transcend context and be used to make broad investment decisions (or better yet, could—once the parameters of automation were established—automate those decisions, displacing human judgment in the process)?[23] To understand the process of translation of default correlation into an immutable mobile, we must begin with the first few pages of Li's article, where a fascinating rhetorical drama unfolds.

The first sentence of "On Default Correlation" is revealing: "The rapidly growing credit derivative market," it reads, "has created a new set of financial instruments which can be used to manage the most important dimensions of financial risk—credit risk" (Li 2000, 2). Immediately, a realist drama unfolds: the scene is chaotic, fast moving, risky. That material reality calls for action, rewarding those who act first and with prime efficiency. As proof of these *facts*, a dynamic, fast-moving, already established market constantly deploys various financial "instruments" useful for managing risk. These instruments, as the remainder of the opening paragraph elaborates, are ubiquitous, and they are, it goes without saying for Li's technical audience, mathematical in nature. At the outset, then, mathematical statements are taken up as objects—they are not formulas or equations or models, they are *instruments* that can help weary investors navigate the "rapidly growing" credit derivative markets. Revealing too is the presumed causal relationship between the market and financial instruments, the former calling forth the latter. Mathematical instruments, Li's discourse suggests, are only deployed out of necessity in order to deal with the increasing complexity of investment risk. This is how a realist drama evolves, positioning Li not as advocate or interpreter but as unmediated witness to the realities of the financial market.[24]

Yet the list of "financial instruments" described in the first paragraph does more than establish a realist frame (a frame rhetorical scholars have shown to be "the default rhetoric" within economics (Aune 2001, 40; Hanan, Ghosh, and Brooks 2014). It also builds a network, linking Li's thinking to a variety of institutionalized financial products, establishing those accepted financial practices as nodes of connection upon which his work will build. "Credit default swaps," "return swaps," "portfolio[s] of credit risk," "equity tranche[s]," "collateralized bond obligation[s] (CBO[s])," "collateralized loan obligations (CLO[s])"— taken together, these signify an already established assemblage of financial instruments/products/practices, all of which orbit around and compose the credit derivative market even as their a priori existence reinforces the integrity of the article's realist style (Li 2000, 2).

As we know, this assemblage of instruments first began to emerge in the late 1970s and early 1980s with the creation of MBSS, at the same time—not by chance—that mathematization began to dominate the financial industry.[25] As we also know, the main constraint on the growth of that assemblage was default correlation or, more precisely, how to accurately predict default correlation between loans within a tranched MBS. As Li attests, "surprising though it may seem, the default correlation has not been well defined and understood in finance" (2000, 2). Having established the epistemological problem, Li's article then elaborates on the existing conventional solution, the "discrete default correlation" approach, an approach limited by the discrete time intervals (usually one year) that it used in order to apply standard correlation formulas within statistics. He calls this method "discrete default correlation" (2000, 3). This approach, Li argues, has several drawbacks: dependence on a particular time interval is blind to empirical studies of credit default behavior over time or with regard to economic cycles; an arbitrary time interval is also impractical, as most investors want a continuous distribution of default risk over time (five years, ten years, thirty years, etc.). But most importantly for his argument, Li asserts,

> default is a time dependent event, and so is default correlation. Let us take the survival time of a human being as an example. The probability of dying within one year for a person aged 50 years today is about 0.6%, but the probability of dying for the same person within 50 years is almost a

sure event. Similarly default correlation is a time dependent quantity. Let us now take the survival times of a couple, both aged 50 years today. The correlation between the two discrete events that each dies within one year is very small. But the correlation between the two discrete events that each dies within 100 years is 1. . . . This paper introduces a few techniques used in survival analysis. These techniques have been widely applied to other areas, such as life contingencies in actuarial science and industry life testing in reliability studies, which are similar to the credit problems we encounter here. (2000, 3)

I cite the article at length here because of the significance of the moment, one that amounts to a radical transformation of the problem of default correlation, offering a first glimpse of the horizon of judgment that will eventually inform the Li copula. The conventional "discrete default correlation" approach relied on a combination of historical default data and Gaussian modeling theory to *approximate* default risk within a discrete time period.[26] Li accepts and builds upon Gaussian modeling theory (more on that later), but he substitutes credit default swap (CDS) data for historical default data. Why does he do this and how does he justify it? The analogy above between mortality probabilities and risk assessment is the key.

While enumerating the many problems with the conventional discrete approach, Li offers a seemingly innocuous example: "Let us take the survival time of a human being." This "example," however, foregrounds a trajectory of analogical reasoning that only becomes clear a few lines later, where the novelty of Li's paper emerges from the application of "techniques used in survival analysis" to the problem of default correlation. If one considers mortgages analogous to people, Li reasons, and default analogous to death, then using actuarial math from that domain becomes a potential avenue for addressing the problem of default correlation. This analogy marks the fulcrum on which Li's copula rests, but how exactly does it transform the problem-situation? Consider, prior to Li's work default correlation marked the boundary of computability in quantitative finance. Conventional approaches were limited because of a healthy fear of the unpredictability of default correlation and the absence of sufficient historical default data to construct reliable models. If we accept Li's analogy between life expectancy and mortgage default, however, everything changes: actuarial science has used increasingly sophisticated mathematical tools to accurately predicted life expectancy correlations for insurance companies for decades. By analogy, if life expectancy correlations are mathematically predictable, then default correlation must also be mathematically predictable. The problem-situation transforms accordingly: default correlation no longer marks the boundary of computability; instead, it names a poorly understood but mathematically computable phenomenon, which in turn transforms it from an intractable constraint on the credit derivatives market to a potential engine for that market's growth.

To better understand exactly how that engine propelled financial markets forward, let's look at Li's analogy more closely, for it is no simple linguistic turn. Instead, the analogy diffracts in multiple directions, authoring a series of translations of the problem-situation called default correlation. Note, for instance, that if one accepts Li's analogy—that mortgages are like humans and default like death—the shift to actuarial math and survival science is quite logical, *but that observation conceals the more significant transformation of mortgage defaults from a scalable into a nonscalable phenomenon*. To elaborate, human longevity, at least currently, is a nonscalable phenomenon, meaning the largest instance of lifespan is not much larger than the average lifespan. Average life expectancy might be 78 years, yet the chances of someone living even twice that long are nil (the oldest age recorded is 122). Mortgage default, on the other hand, is a scalable phenomenon, meaning the largest instance of the phenomenon can be far larger than the average (Taleb 2007). Recall that in 2000 the main obstacle to predicting default correlation was the limited amount of historical default data that existed, indicating the average rate of default was historically low (which it was and is—well below 5 percent). Yet it is obviously possible for the mortgage default rate to double, triple, even quadruple the average. In 2004, for instance, the default rate on single-family mortgages was 1.39 percent, but by 2010 it was 11.27 percent, or over eight times higher (Stlouisfed.org).

Li's analogy, however, is making a multipronged argument: (1) that we have misunderstood mortgage default, (2) that it may not be the scalable phenomenon we think it is, and (3) that an overreliance on historical default data has distorted our perception.[27] Using historical default data one can easily "prove" (as above) that default correlation is scalable, and, therefore, highly unpredictable. Yet that "proof" relies on scant data, possibly falsifying and certainly weakening any argument dependent on it. Li's analogy compels his readers to reconsider the value of historical default data: perhaps default correlation is not scalable after all; perhaps historical default data *are* misleading. In so doing, historical default data morph from a valuable empirical resource for understanding default correlation into, at best, a constraint on predictive models and, at worst, a distortion of the very phenomenon Quants of structured finance are trying to understand. But what if there is a better alternative, a resource that would provide a wealth of data to both clarify our understanding and improve the reliability of any predictive models of default correlation?

Enter credit default swaps (CDSS). The analogy between default and death granted Li access to all the mathematical tools of actuarial science, but without sufficient data no algorithm, no matter how sophisticated, could model default correlations accurately. Li needed more data, and his great innovation was to replace historical default data (whose ethos he had already undermined) with data from the CDS market. CDSS are the counterpart to traditional loans: with traditional loans a lender makes money off the interest on the loan, but with CDSS one offers insurance against default on particular loans. If prices of a CDS increase, the probability of default increases. Li reasoned that if the CDS prices of two loans increased and decreased together that indicated they were highly correlated. This was the last major puzzle piece: Li had access to actuarial science, and now he had the necessary data to build and test his copula.

Mathematical Statements as Principles of Composition

We are now fully embedded in a hybrid—an admixture of the real and the imaginary, the actual and the virtual, the human and the nonhuman—a place where market risk is immanent, default correlation is poorly understood but calculable, mortgage defaults and life expectancy are akin, historical defaults are not to be trusted, and credit default swaps are a rich and deep resource of much-needed default data. The problem-situation called default correlation has been deconstructed and many of its component parts transformed. Now it must be recomposed. For that task Li turns to mathematics as a powerful principle of composition—a means of rendering out of analogy and conjecture something hard, something tractable, something immutable.

But how does math actively compose a world? How can we understand it as a constitutive and not merely reflective force, one that may even possess its own forms of agency? Perhaps a closer look at Li's mathematical discourse will offer some insight. As is common among mathematicians, Li begins with abstraction, inventing a "random variable called 'time-until-default'" and thereby transporting both himself and his reader from the empirical realm of the particular and historical to the mathematical realm of the abstract and general (Li 2000, 3–4). This is a key rhetorical moment in most mathematical discourse, for it enables one to render commensurable what was initially incommensurable. In this case, Li wants to treat all debt obligations and all defaults as commensurable, so he creates a *formal* category "A" to represent all debts and then combines that with his "time-until-default" variable to form a new hybrid, "T_A," which represents the time until default of any and all debt obligations (now rendered equivalent through "A"). Li then represents the probability distribution (simply the points on a graph of the likelihood of default at different times) of A's default with

$$F(t) = Pr(T_A \leq t), t \geq 0$$

and the inverse, A's "survival," with

$$S(t) = 1 - F(t) = Pr(T_A > t), t \geq 0$$

Already apparent here is the way mathematical discourse enables one to "obtain *nth* order forms that are combined with other *nth* order forms coming from completely different regions" (Latour 1987, 244). Those nth-order forms are part of the compositional power of mathematics. In fact, it is the emphasis on abstract form that renders what was once intractably particular—the diversity of loans and their individual potential for default—into a commensurable and thus computable form.

Yet something more is afoot in these initial mathematical moves. For inside every mathematical code stands the central ideas for which it is the vehicle. And the central ideas within the functions $F(t)$ and $S(t)$ are

1. that loans are like people and default like death;
2. that actuarial science can therefore be used to address default;
3. that default, through actuarial science, is best expressed as a probability function;
4. that, since default is nonscalable (just like human longevity), one can use Gaussian bell-curved probability theory to effectively model default risk over time; and
5. that, as equivalent and commensurable probability functions, the individual particularities of each loan are unnecessary for assessing default risk.

By critically reading Li's mathematical discourse in this way, a more detailed map of the many judgments informing the Li copula begins to emerge, and as that process evolves we will see how the judgments baked into that mathematical discourse slowly displace the forms of practical judgment that guided financial decision-making prior to Li's work. Before the copula, for instance, practical judgment encouraged diversification as a heuristic to avoid exposure to default correlation (McLean and Nocera 2011; Krugman 2009). That heuristic required attention to and evaluation of the particular debt obligations within a tranched mortgage-backed security. Li's actuarial approach, however, transports us to a world where particular mortgage defaults no longer exist—at least not in the same form. They have been transformed into commensurable probability curves, a transformation that promises to reveal the hidden patterns of default correlation while simultaneously rendering those old forms of *phronesis* (diversification and direct assessment) obsolete. Having set the foundations for his mathematical treatment of default correlation, Li is now in a position to address more complicated default situations and complete his recomposition of the credit derivatives market.

With the basic concepts and functions established for default risk of a single loan, Li still needed to develop a mathematical way to represent the more complicated problem of default correlation between loans. Predicting the default risk of a single loan is a challenging but not impossible problem to accurately approximate. Much more difficult are dependent or conditional probability problems, ones that ask, "If mortgage A defaults, what is the probability of mortgage B defaulting at time X." Here again, however, Li can fall back on his original analogy, which sponsors all kinds of transversal connections between default and survival science. These transversal connections come in the form of algorithms designed by insurance actuaries to determine survival probabilities of one spouse, for example, if the other spouse dies. Transversally stretched into the realm of default correlation, these more complex actuarial algorithms, such as the "hazard rate function," the "joint distribution function," and the "survival function," are combined and substituted into Li's original functions F(t) and S(t), such that we move from the original "survival function"

$$S(t) = 1 - F(t) = Pr(T_A > t), t \geq 0$$

to its expression "in terms of the hazard rate function"

$$S(t) = e^{-\int_0^t h(s)ds}$$

to "the default correlation of two entities A and B" (Li 2000, 6–8).

$$\rho_{AB} = \frac{E(T_A T_B) - E(T_A)E(T_B)}{\sqrt{Var(T_A)Var(T_B)}}$$

One can easily become bewildered by these mathematical transformations, but it is not necessary to know the calculus of survival science to see that with each step there is a slow scaling up from the basic functions that accounted for the default risk of a single loan to "the comparison of groups of individuals" and their relative survival probabilities to "more complicated situations, such as where there is censoring or there are several types of default" (Li 2000, 7). This process of scaling up does two things: first, it strengthens the original simple functions $F(t)$ and $S(t)$ by association with other more complicated and well-established functions in survival science; second, each mathematical association and substitution simultaneously improves the tractability of those original functions, increasing their ability to be used for more and more complex situations (mortgage pools, for example, or defaults between different kinds of debt obligations, or even externally imposed censoring).

The remainder of Li's article leads the reader inexorably toward his ultimate equation. We are taken through his construction of credit curves based on the credit default swap data mentioned earlier, his analysis of other means of creating credit curves and their comparative weaknesses, and his introduction of mathematical copulas and their uses within actuarial science, all of which eventually brings us to his final "bivariate normal copula function" (Li 2000, 16):

$$Pr[T_A < 1, T_B < 1] = \phi_2(\phi^{-1}(F_A(1)), \phi^{-1}(F_B(1)), \gamma)$$

The left side of this equation establishes a "joint default probability" function between loan A and loan B (Li 2000, 16). The right side of the equation then reexpresses that joint default probability as a copula function (Φ) that reduces default correlation to a single constant (γ). By the time we arrive at this algorithm, however, all the assumptions and delimitations with which Li's paper began are barely visible, obscured by the numerous mathematical steps taken— each of which is irrefutably "true" within the world the article constructs. The result is that, by the end of the article, the horizon of judgment and domain of validity informing the Li copula have almost completely vanished, buried in a series of *correct* mathematical steps.

This is how immutable mobiles are built. Through invention, substitution, and transversal connections an algorithm is assembled. That process of assembly transports us from the local (the singular mortgage) to the global (the universal debt-obligation-default-probability-curve), with the ultimate purpose being to render the original algorithm into a universal computation machine. Why? So that the resulting algorithm is powerful enough to function as a principle of composition—a "technology" that can gather, translate, reconfigure, and ultimately transform the semiotic-material world. Mathematical algorithms are not only about solving problems; they are also about enabling new possibilities—new hybrids—to emerge. Having traced the discursive practices that Li used to assemble the copula, we can now turn toward those hybrids, attending to the copula's material manifestations and its rapid recomposition of global finance.

Terraforming

The previous analysis examined the entanglement of traditional and mathematical strategies of argument within Li's article in order to see how those strategies translated and transformed the problem-situation called default correlation. How did the resulting algorithm materially impact global finance? Much is at stake in this question, and addressing it promises to help us not only see the materiality of the Li copula but understand more broadly the terraforming modalities of mathematical discourse.[28] By way of entry, let's begin, as Latour suggests, with the *vinculum*—the new relations that the Li copula made possible within structured finance (1993, 129).

Perhaps the most significant relation to emerge from the Li copula can simply be called "the algorithmic relation." Algorithmic relations are nothing new within finance or any number of other sectors of society. When you use the Google search engine, you enter into an algorithmic relation. When you use a mapping app like Ways, you enter into an algorithmic relation. When you search Netflix for new shows, you enter into an algorithmic relation. Yet in each case those algorithms are powerful for the same reason—they are able to interrupt a space of practical judgment and human agency and replace it with algorithmic automation (Finn 2017, 20). You no longer need to engage your parahippocampus in the same way in order to navigate a city (Burgess, Maguire, and O'Keefe 2002; Sparrow, Liu, and Wegner 2011), nor do you need to read descriptions of particular films to find something to watch (Amatriain 2012;

Madrigal 2014). The algorithm's power lies in its capacity to displace (not eliminate but decenter and marginalize) those forms of human agency and judgment. The Li copula is no different in this regard. It effectively interrupted a space of human agency and judgment, displaced it with a mathematical algorithm (within its domain of validity), expanded the realm of computability within quantitative finance, and provided a precise measurement of default correlation, conveniently reduced to a single constant. Evidence of its perceived veracity and tractability can be found in many locales: the copula circulated quickly within the intellectual community of quantitative finance; Li's status within that community grew rapidly, evidenced by both his climb up the corporate ladder and citation indexes within his field; and, importantly, the major rating agencies (Moody's, Fitch, etc.), whose job was to rate investment tranches, adopted the copula as the primary means for determining default correlation and thus risk exposure of particular tranches of pooled securities (Coval, Jurek, and Stafford 2009; Lewis 2010). The use of the Li copula as the primary means to rate default risk allowed the algorithmic relation it introduced to spread, consolidating and displacing previous realms of human judgment, thereby increasing its power and influence, and ultimately expanding credit derivative markets in unprecedented ways.

The CDO and its derivative the CDO^2 are perhaps the best examples of the Li copula's material recomposition of global finance.[29] CDOs are similar to mortgage-backed securities in that a securities firm pools and tranches debt obligations into structured investments. The main difference is that instead of pooling mortgages, CDOs draw from multiple sectors of the economy—everything from manufacturing debt to credit card debt. Invented in 1987, CDOs were thought to be relatively safe because they included diverse forms of debt from different economic sectors, but just as with MBSs the inability to accurately assess default correlation severely limited the CDO market (even more so than the MBS market due to the seemingly impossible problem of predicting default correlations between radically different forms of debt). As a result, the CDO market was 69 billion in 2000, a mere fraction of the credit derivatives market at the time (McLean and Nocera 2011, 123). The Li copula effectively erased those differences, rendering all debt obligations commensurable and then using credit default swap data (which existed for most major forms of debt) to construct their default probability curves. Suddenly, one could use Li's new algorithm to assess default correlation without attending to the underlying loans or their differences, for all loans were commensurable in the mathematical realm of the copula.

The Li copula, however, did not simply accelerate or extend the fledgling CDO market, it enabled new products—new hybrids—to flood that market. The basic idea behind pooling and tranching, whether into an MBS or a CDO or anything else, is to create investment products that are safer than the individual loans within the pool. Consider a simple two-loan pool. In this hypothetical pool each loan has a default risk of 10 percent and those default risks are uncorrelated. If we pool and tranche those loans such that the junior tranche pays \$1 if neither loan defaults but \$0 if either or both default and the senior tranche pays \$1 in every scenario except if both loans default, then the risk of default for the senior tranche is reduced from 10 percent to 1 percent (since it is protected by the junior tranche).[30] Individually, the loans would be rated as BB+ and thus could not be sold to government-sponsored institutions like Freddy Mac or Fannie Mae, but once they are pooled and tranched, the senior tranche would receive an AA rating, which is well within the parameters of investment-grade bonds (Coval, Jurek, and Stafford 2009, 8–9). However, in our hypothetical we assumed no default correlation between our two loans. If default correlation increases even marginally, we will get radically different risk exposures for our junior and senior tranches, which is precisely the problem that the Li copula purported to solve. Armed with a mathematically precise measurement of default correlation, securities firms were able to create a variety of new financial products. They took those below-investment-grade tranches from different CDOs and MBSs, pooled them together, tranched, and, using the same logic as above, produced CDO^2s that were well above investment grade, something they could easily do because the rating agencies implemented the same copula to rate the tranches as they did to create them.[31]

These hybrids of the Li copula and the algorithmic relation it introduced completely transformed global finance. As the CDO market started to expand, the old division between mortgages and other forms of debt dissolved, as did the separation between the MBS market and the CDO market. From 2000 to 2005 the CDO market grew from 69 billion to over 500 billion, and by 2004 over half of the underlying collateral of CDOs was made up of mortgage-backed securities (Financial Crisis Inquiry Commission 2011, 130; McLean and Nocera 2011, 201). From 2003 until the financial crisis, the CDO market was the fastest-growing sector within structured finance; in 2006 alone sales for CDOs were 500 billion—as much as the whole CDO market the year before—and with increasing international interest the market grew to over 2 trillion, just shy of the GDP of China in the same year (McLean and Nocera 2011, 123; Benmelech and Dlugosz 2009; International Monetary Fund 2017).

These are not statistics—they are material formations, aggregates of the algorithmic relation introduced through the Li copula and circulated through its broad adoption by both sides of structured finance (the ratings side and the investment side). These numbers are traces of the terraforming power of the Li copula—the way it translated the problem of default correlation, moved the border of computability in the process, introduced an algorithmic relation that automated risk assessment, and eventually gave rise to a variety of new hybrids that enabled securities firms to do various forms of "ratings arbitrage," selling tranches that were initially below investment grade as AAA securities. Like Archimedes's ratios, the Li copula introduced a novel relation into the social collective, out of which grew a variety of new hybrids that Li himself could not have anticipated. Like King Hiero's siege engines, those hybrids radically transformed power relations—in this case, the power relations between global finance, the government structures designed to regulate its many forms, and the citizens that both are meant to serve.

Aftermath and Implications

A decade after the rise and fall of the CDO markets the narrative of human agency, greed, hubris, and ignorance that led to the financial crisis is well worn. In the *Financial Crisis Inquiry Report* of 2011, the authors begin with the conclusions they drew from their multiyear study. The first line reads, "We conclude this financial crisis was avoidable. The crisis was the result of human action and inaction, not of Mother Nature or computer models gone haywire" (xvii). Michael Lewis's *The Big Short: Inside the Doomsday Machine* (an account converted into a major Hollywood film in 2015) placed the blame at the feet of the securities firms and rating agencies, the first of which created "dishonest" CDO investments and the second of which granted those investments artificially inflated ratings because of the "fat fees" received from Wall Street firms like Goldman Sachs (Lewis 2010, 73). Nobel laureate Paul Krugman blamed "politicians and government officials," who "should have realized that they were re-creating the kind of financial vulnerability that made the Great Depression possible." Krugman characterized the behavior before the financial crisis as "malign neglect" (Krugman 2009, 155). Others claimed the spread of subprime mortgages throughout the global financial system was due to the perverse incentives and pressures created by the CDO markets, which rewarded banks for creating more subprime mortgages (Morgenson and Rosner 2011, 283). As divergent as these narratives appear, one finds the same antagonist and the same purpose in each: in every account human agency controls the drama—whether in the form of greed, or ignorance, or "malign neglect"—with the ultimate purpose of assigning blame. And while many of these narratives are accurate and insightful, they are also partial—human-centric in a way that obscures mathematical discourse as a principle of composition that actively remakes and expands our social collectives.

We have already seen how the Li copula translated the problem-situation called default correlation within quantitative finance, introducing through a central analogy and a series of transversal connections a novel algorithmic relation that materially transformed structured finance on a global scale. Now we can examine the economic fallout, not in order to shift blame from human actors onto algorithms but rather to better understand the translative powers of mathematical discourse (in this case within the context of structured finance). Only then will we be in a position to step back and consider the implications of this study for rhetoric and mathematics more broadly.

What role did the Li copula have in the economic collapse? In terms of implementation, it is fairly easy to see how the Li copula enabled the spread of subprime mortgages. Equipped with a mathematically precise way to determine default correlation, the community we call structured finance no longer felt the need to investigate the underlying collateral of tranched securities. Combine that mentality with the hierarchical payout structure that tranching provided, and you have the recipe for the spread of subprime mortgages. Take all those low-rated MBS tranches, for instance, repool and retranche them, then rerate them using the Li copula, and you can create new highly rated securities. Your starting point for that retranching process, however, is radically different: instead of a pool with 5–10 percent subprime collateral (the historical average for MBSS), you now start with 40–70 percent subprime collateral, yet you have a mathematical means to sell 70–80 percent of that pool as investment grade (Lewis 2010, 73). Practical judgment tells you that eventually those subprime mortgages must trickle upward into the tranches being sold to government-backed financial institutions. But with the dominance of the algorithmic relation—and the apotheosis of algorithms generally within modern economics—few actually looked at the underlying collateral, which enabled subprime mortgages to infiltrate global financial networks at almost every level.

Even this narrative, however, remains too human-centric and too instrumentalist for us to see mathematical discourse as a principle of composition. To escape the gravity of these paradigmatic narrative forms and begin to grasp mathematical discourse not merely as *reflective* of reality but also as *productive* of reality, we must consider the deeper structures within the Li copula. We can begin to do that by returning to its central analogy and the difference between scalable and nonscalable systems. Recall that implicit in the analogy is an argument that mortgage default, like human longevity, is a nonscalable system. To support this claim Li used Gaussian probability theory and credit default swap data to construct credit curves for probability of default over time, thereby "demonstrating" the nonscalable nature of default. What went wrong? CDS data, while plentiful, turned out to be fairly misleading. The CDS market was not created until 1994, meaning the models were dependent on six years of economic data. During that time housing prices nationally only increased, which artificially decreased default correlations between CDSs. Coval, Jurek, and Stafford's (2009) work shows that even a modest decrease in housing prices exploded the default correlations within the CDS market, suggesting that *when housing prices decrease, default correlations increase in a scalable manner*. The substitution of CDS data for historical default data thus introduces into the copula's horizon of judgment another assumption—that housing prices will increase. Yet this assumption is curiously nonhuman, a feature of CDS data itself and not the consequence of the mathematician "behind" the copula.

Regardless of authorship (or its absence), the difference between scalable and nonscalable systems has serious consequences. Nonscalable systems can be effectively modeled using Gaussian probability theory (bell curves and normal distributions) because there is a built-in stability to these systems that keeps things "normally distributed," that is, clustered around the mean with relatively few outliers. Gaussian models for scalable systems, however, are extremely error prone because, as Nassim Taleb points out, "the bell curve ignores large deviations, cannot handle them, yet makes us confident that we have tamed uncertainty" (Taleb 2007, xxix; Haug and Taleb 2011). Unfortunately, the problem does not end there: Gaussian models not only predict scalable systems poorly; they can actually conceal scalability even as they increase risk exposure. Outside of highly stable, normally distributed systems (such as life expectancy), correlation measures are fairly meaningless as a means of prediction due to their instability, "yet people talk about correlation as if it were something real, making it tangible, investing it with a physical property, reifying it" (Taleb 2007, 239).

The problem here is not with correlation per se but with the reification of correlation into a "matter of fact," into something "hard" that one can use to make financial decisions. This is one of the roles that mathematical discourse plays in Li's article—it helps harden his analogical conjecture into an algorithm with apparent mathematical rigor. Within the world Li's analogy invents, his mathematics *is rigorous*. Yet outside that world, in the world of scalable financial markets and default correlations, the reifying rigor of Gaussian probability curves conceals risk for a simple reason: in normal distribution probability theory there is an exponential decline in the odds of an outlier as one moves farther from the center. Put differently, when one uses a Gaussian model to forecast one implicitly accepts the assumption that as one moves away from the average the likelihood of an event becomes *exponentially rarer, crashing faster and faster toward zero.* When one develops mathematical models based on Gaussian mathematics, then, the judgment that the system one is modeling is "normal" (nonscalable) is already baked into the mathematical structure, and the system's scalable risk is effectively obscured by the sophisticated mathematical models deployed.

Note the limits of both human-centric and instrumental narratives here: Li is not an all-powerful mathematical wizard, nor the unwitting architect of the subprime mortgage crisis. Instead, he is a situated actor, and when he speaks, whole discourses of power speak. He did not invent Gaussian probability theory, nor did he invent the quantification-of-finance movement that primed his intellectual community for algorithmic automation. These are broader discursive formations within which Li was caught and for which he was a spokesperson. He simply used widely accepted practices within quantitative finance and actuarial science to expand and test a thought experiment. That process *brought him* to the Li copula. That copula is neither Li's invention nor his discovery; it is an apparatus that sliced the world of structured finance anew, introducing a novel relation, giving rise to myriad new hybrids, and transforming our economic structures accordingly.[32]

As an apparatus, the Li copula possesses its own unique ontological DNA, rearranging through various translations the hierarchy of evidentiary value, rendering that which was beyond the realm of calculation suddenly calculable, introducing new relations and novel hybrids, and thereby transforming the subjectivity of its users. As Finn observes, "implementation runs both ways—every culture machine [algorithm] we build to interface with the embodied world of human materiality also reconfigures that embodied space, altering cognitive and cultural practices" (Finn 2017, 49). How did the Li copula alter our cognitive and cultural practices? Beyond the commensurabilities it constructed and the dissociations it encouraged between investors and investments, beyond the algorithmic relation it introduced that enabled human actors to construct new human-nonhuman hybrids, consider the way the copula reduced default correlation between untold numbers of loans to a single constant (γ). This feature of the copula, we must note at the outset, is not Li's—it is characteristic of mathematical copulas themselves. Yet that aggregation of default correlation into a constant had several cognitive and cultural consequences: it further reified the copula into a *technology* (instead of a model) for *measuring* (instead of estimating), it increased the perceived applicability of the copula across investment domains, and, most critically, it provided an anchor that promoted an image of stability, encouraging users to underestimate risk.

There are many ways in which numbers can influence one's perception: the simplest example emerges from an experiment in which Amos Tversky and Daniel Kahneman—who first studied the anchoring effects of numbers—asked participants to guess "the percentage of African countries in the UN." One group spun a "wheel of fortune" rigged to stop at ten, then guessed the percentage. Another group spun the same wheel rigged to stop at sixty-five. The results were surprising: the first group's average guess was 25 percent and the second group's was 45 percent. Anchoring, Kahneman explains, "occurs when people consider a particular value for an unknown quantity before estimating that quantity" (Kahneman 2011, 119).

While manifest in many contexts, the role of anchoring in what Tversky and Kahneman call "disjunctive events" clarifies the copula's agential force. Disjunctive events take place within complex systems (such as nuclear reactors or human bodies) where many elements depend on each other in order for the system to function. Within these systems, even though "the likelihood of failure in each component is slight, the probability of an overall failure can be high if many components are involved" (Tversky and Kahneman 2011, 428). With the network the Li copula helped produce there were literally thousands, perhaps millions, of components involved (CDS data, pooled debt obligations, Gaussian probability curves, CDO hybrids, securities firms, rating agencies; the list goes on), yet the risk of that complex network, which the Li copula exponentially expanded, was gathered and condensed into a single number—a constant that belied the scalable risk of disjunctive events. That constant exacerbated the funnel-like structure of the Li copula, which homogenized the field of judgment that informed decision-making within structured finance and in so doing rendered that sector simultaneously larger and more fragile; that is, larger but at the same time increasingly funneled into a single node of judgment parameters that, if violated, would lead to cataclysmic collapse. In this way, one can see that the copula did not simply "represent" a series of a priori relations within the credit derivatives market and distill them into mathematical form. Instead, it actively gathered the prior network and concentrated it such that it relied almost solely on the empirical veracity of CDS data. Because the CDS market was itself extremely scalable, the copula effectively increased structured finance's exposure to scalable change. As the copula was deployed more and more broadly, its anchoring effect diminished the perception of risk even as its funnel-like structure rendered the structured finance network larger, more homogeneous, and more unstable.[33]

These are the ways in which mathematical discourse can alter our cognitive and cultural practices, yet to see these discursive-material entanglements we must extend our understanding of both rhetoric and math. Within the configuration that animates this analysis, for example, rhetoric expands beyond argumentation and symbolic action without leaving them behind; rhetoric extends to include the study of symbolic-material relations, their emergence, their productivities, and their material consequences, adding to our accounts of words and deeds what Nathan Stormer describes as studies of "addressivity," or the ways symbolic-material practices establish "a set of capacities for address that forms and fades within fields of power" (Stormer 2016, 306).[34] One's scholarly positionality shifts accordingly from the confines of negative critique—Who did this? Who's responsible? or even How did math betray us with obfuscation and reductionism?—to the realms of symbolic-material production, where we ask instead—How do symbolic-material practices make certain forms of addressivity possible; How does math as a symbolic-material practice translate, transform, and mediate, enabling novel relations to emerge that expand our social collectives; What forms of agency must we account for in order to understand the unparalleled productivity of mathematical discourse; and How can we trace the ontological force of algorithmic relations and the ways they shift our capacities for address?

These questions implicitly reject old modernist divisions between rhetoric and math, humans and nonhumans, societies and natures, not out of some antipathy for the past but because those ontological relations are becoming increasingly inadequate for understanding the proliferating realities of the twenty-first century. Instead, this essay attempts to practice methodological symmetry, which requires one to attend to the multidirectional feedback loops between rhetoric and math, humans and nonhumans, societies and natures, such that those divisions begin to transform into hybrid networks and we begin to see their productive entanglements.[35] The exemplar in this essay is the Li copula and its horizon of judgment, which is both human and nonhuman, individual and collective, rhetorical and mathematical, yet powerful in the way it displaced the practical forms of judgment that preceded it, introduced a new algorithmic relation, and authored a transformation of global finance that eventually collapsed in on itself.

Despite that collapse, the purpose of this analysis is not to critique structured finance for being duped by an algorithm. The purpose is to offer an alternative to that epistemological (often positivist) critique, tracing instead the ways traditional rhetorical modalities (like analogy and argument) entwine with the rhetorical modalities of numeracy (abstraction, substitution, commensurability, etc.) to generate new relations between humans and nonhumans. Those new relations are the entities that mark the irreversible entanglement of symbolicity and materiality, the results of which often expand our social collectives. This analytic approach, and the hybrid analytic it embraces, seeks not only to reveal the ways mathematical discourse can displace practical judgment but, perhaps more importantly, to promote critical engagement with practices of mathematization and algorithmic automation, with the hope of encouraging others to become critical informants—citizen-scholars with the skills to unpack the algorithmic black boxes within fields of power (whether economic or otherwise) and trace the ways they translate and expand our social-material worlds. Only in this way will rhetoric as a field be in a position to respond and contribute something of value to an increasingly mathematized and algorithmically driven world.

Notes

1. Too many colleagues to name have helped me improve, refine, and sharpen this argument. Any remaining errors are mine alone. Special thanks to James Wynn, Joshua Hanan, Kundai Chirindo, and David Schulz, whose generous feedback strengthened the manuscript considerably. An extension of this essay can be found in Reyes 2019. While research on the rhetoric of economics is too extensive to cite in full, a few touchstones for this essay include McCloskey (1998), Mirowski (1988), Mirowski (1991), Aune (2001), Houck (2001), Goodnight and Green (2010), Hanan, Ghosh, and Brooks (2014), and Hanan and Hayward (2014). The last volume offers an extensive bibliography of relevant contemporary research.

2. Many, like Simonson (2014), have noted the "logophilia" of contemporary rhetorical research.

3. As Hanan, Ghosh, and Brooks (2014) show, neoclassical economics relies predominantly on "a deductivist mathematical method" so widespread that it is nearly impossible to be published in top-tier economics journals without mathematical backing (141).

4. I use "harden" throughout the essay in Bruno Latour's (1987) sense to describe how mathematical discourse extends the networks that support particular statements within economics.

5. I will develop the concept of the horizon of judgment in the body of the essay. In short, it refers to the network of human and nonhuman assumptions, constraints, and delimitations necessary for a mathematical proposition to have any specificity or validity. I use the metaphor of horizon to underscore the distant but no less present domain of validity that constrains every mathematical proposition.

6. The question of agency and its locus (or multiple loci) will be addressed throughout the essay. As the argument develops, we will see how human and nonhuman agencies intertwine in the slow formation and articulation of the Li copula. Thus the horizon of judgment that animates the Li copula is no simple distillation of human judgment (Li's or others), but instead an entanglement of human, mathematical, and nonhuman agencies.

7. As will be discussed in the analysis, the discursive shift from "model" and "forecasting" to "mechanism" and "measurement" underwrites a shift from a fallibilist to a realist epistemology.

8. The notion of rhetoric as a multiplicity (rhetorics) rather than a singularity as well as entailing the study of not just symbolic action but the entanglements of symbolicity and materiality that compose our worlds is very much in line with new materialist theories of rhetoric—especially Nathan Stormer's (2016; 2004) notion of rhetoric as polythetic.

9. A fuller explication of a constitutive approach to mathematical discourse can be found in Rotman's work (1993; 1987) as well as in Mudry (2009) and Reyes (2004; 2014). Maurice Charland's work (1987) offers an elaboration of the theory of constitutive rhetoric, and Karen Barad's excellent book *Meeting the Universe Halfway* shows how a constitutive approach challenges several intellectual orthodoxies, including the central tenets of representationalism—that words and things are both independent and determinate, that the world is composed of individual entities with definite boundaries and characteristics, and that proper experimental measurement reveals the intrinsic properties of independently existing objects (2007, 107).

10. Latour's work has been taken up in a variety of ways in rhetorical studies, though to my knowledge no rhetorical scholar has addressed his work on mathematical discourse. For those interested in potential intersections between rhetoric and Latour, Lynch and Rivers's (2015) edited collection is an excellent starting point.

11. Quotations from Plutarch, *Life of Marcellus*, are from Plutarch (1968, 7–9).

12. For economy's sake, I dare not go too far into the woods on this point. Suffice it to say that whole books have been written teasing out the implications of mathematical realism on everything from mathematics itself to mathematics pedagogy to the ontological positioning of the human vis-à-vis nature to the relationship between democracy and the mathematical and natural sciences. For those interested in pursuing such topics, see Rotman (1987; 1993; 2000), MacKenzie (2006), Latour (1987; 1988; 1999), and Barad (2007).

13. This point aligns with Joshua Prenosil's argument that Latour's work opens the way for "a rhetoric of inartistic proofs . . . keenly interested in the way that power is remediated in contemporary practice" (2015, 98). In this case, the "inartistic proof" is the fabricated algorithmic technology that materially transforms social-material reality.

14. Latour's position here is consistent with his broader philosophical rejection of modernist metaphysics, which emerged historically with the rise of Cartesianism, modern algebra, and a renewed commitment to mathematical realism (Cifoletti 2006; Latour 1993; 1988).

15. Reduction of incommensurability to commensurability, heterogeneity to homogeneity, is only the first moment of mathematization and if emphasized too much can obscure the fact that the purpose of such reductionism is not sameness but transformation, differentiation, and extension. Through a reduction to form, math can reveal potential relations that often materially transform networks. Like an atomic detonation, which must first implode before exploding, math first reduces or, better yet, transforms into a common form the elements and constraints of the problem-situation, which occasionally leads to the realization of a novel relation that (again, occasionally) becomes materially manifest in the world.

16. The question of what becomes of rhetoric through a Latourian approach would require a whole other essay to adequately address. Suffice it to say that for all his strengths, this question reveals a weakness in Latour's thought, for he marshals a fairly conventional notion of rhetoric as argument and persuasion even as his novel approach seems to call for an equally novel notion of rhetoric (Latour 1987). A burgeoning scholarly conversation is currently emerging around this very question, and at this point I find Stormer's notion of rhetoric as polythetic the most developed response (2016, 300). Paul Lynch's and Nathaniel Rivers's recent edited volume *Thinking with Bruno Latour in Rhetoric and Composition* (2015) is an excellent point of entry into this conversation.

17. This approach is both pragmatic and philosophically disinclined to adopt a single method, for the point of analysis here is not merely to reveal the hidden (though this is a crucial element of analysis) but more importantly to multiply our critical capacities for address and encounter with mathematical discourse, whether within economics as a field of power or within other domains. Latour writes about analysis as a force of multiplication (2005, 144).

18. For more on the impact of MBSS on the rise of neoliberalism and the transformation of the home into an "abstract financial equation," see MacKenzie (2006) and Hanan (2010, 184).

19. The risk changes because investors in MBSS are paid out hierarchically. For example, AAA investors would only be at risk of losing money if more than 20 percent of a mortgage pool defaulted, while that percentage would decrease with each tranche subordinate to the AAA tranche.

20. Hanan, Ghosh, and Brooks's "Banking on the Present" offers an excellent overview of the rise of MBSS within a broader critique of neoclassical economics. For more on mortgage-backed securities and their relationship to subprime mortgages and the 2007 economic collapse, see Coval, Jurek, and Stafford (2009), Fabozzi, Bhattacharya, and Berliner (2011), and McLean and Nocera (2011).

21. The basic mathematics behind default correlation is as follows: correlation is considered a probability function of the relationship between two debt obligations, meaning it ranges from 1 to −1, where .2 default correlation would represent a 20 percent probability that if one mortgage defaults the other will as well. In contrast, a −.2 default correlation would represent a negative 20 percent probability that if one mortgage defaults the other will default, meaning that if one defaults the chances of the other defaulting as well decreases. For more background on problems of default correlation, see Coval, Jurek, and Stafford (2009).

22. As Coval, Jurek, and Stafford note, the use of the Li copula to "repackage risks and to create 'safe' assets from otherwise risky collateral led to a dramatic expansion in the issuance of structured securities, most of which were viewed by investors to be virtually risk-free and certified as such by the rating agencies" (2009, 3). Salmon elaborates: "using Li's copula approach meant that ratings agencies like Moody's—or anybody wanting to model the risk

of a tranche—no longer needed to puzzle over the underlying securities. All they needed was that correlation number, and out would come a rating telling them how safe or risky the tranche was" (2009, 19).

23. Human judgment has long been associated with the capacities of taking in particular information from one's context and making decisions based on a combination of that information and one's prior experience. This is the classical notion of human judgment one finds elaborated in Aristotle's notion of *phronesis* (Aristotle 1981; Miller 2003). Algorithmic automation displaces those capacities with abstract equations presumed to be free of subjective human judgment. As I will elaborate later in the analysis, however, algorithmic automation is neither free of human judgment nor merely a means of spreading one particular set of human judgments. Instead, it is an amalgam of both human and nonhuman agencies.

24. A basic tenant of realism is that in order to have unmediated access to "the real" one must cast off all ideology and all artifice. Economic realists see the language of mathematics as the best means for casting off artifice, exposing ideology, and discovering the real (Maki 1988; McCloskey 1988). Thus realists describe the world in plain language (sans artifice), positioning the author as witness to the world so described. Realism as a rhetorical style is well developed in Robert Hariman's book *Political Style* (1995) as well as in Reyes (2006). Rhetorical realism in economics is dealt with in Aune (2001). Li's article provides ample evidence of a realist style (2000, 3–5).

25. While the mathematization of economics first began with the rise of probability and statistics in the 1940s and 1950s, it was not until the 1970s that it became a significant force within structured finance, concomitant with the spread of computers and increasing computational power (McCloskey 1998, 112).

26. Several mathematical formulas exist to do this, but the Black-Scholes-Merton formula is the most famous, receiving the Nobel Prize in 1997 for "showing that it is in fact not necessary to use any risk premium when valuing an option" (Royal Swedish Academy of Sciences, October 14, 1997, http://www.nobelprize.org/nobel_prizes/economic-sciences /laureates/1997/press.html). This shift was crucial to convincing many that a mathematically deterministic prediction of option pricing was possible.

27. Some of these elements of argument are explicit in Li's text (such as the collective poor understanding of default correlation), and some are implicit (such as the scalability or nonscalability of default correlation) but nevertheless necessary for the integrity of Li's mathematical treatment.

28. By "terraforming," I mean the ways math transforms our social-material worlds.

29. The Li copula sponsored many other hybrids within the CDO market, such as cash, synthetic, and hybrid CDOs. For details on all the financial products that emerged in the wake of the Li copula, see McLean and Nocera (2011).

30. For a more in-depth explanation of pooling and tranching, see Coval, Jurek, and Stafford (2009, 6–7).

31. By way of example, take the junior tranche in our hypothetical loan pool and imagine combining it with other junior tranches from other CDOs or MBSS. Using the same logic, one could pool those debt obligations and tranche them, creating CDO2 products with senior tranches that would be above investment grade. With each repooling and retranching process, however, the number of risky underlying debt obligations (such as subprime mortgages) increases, yet the Li copula convinced many within finance that the particular characteristics of the underlying loans were irrelevant. As *The Financial Crisis Inquiry Report* (2011) later discovered, "approximately 80% of these CDO tranches would be rated triple A despite the fact that they generally comprised the lower-rated tranches of mortgage-backed securities" (127).

32. I borrow the term "sliced" from Barad (2007), who uses it to describe the power of new knowledge as the power to slice social-material relations into novel configurations.

33. Anchoring is only one dimension of the copula's agential force, and it would not have had the impact it did in isolation. The copula's reduction of default correlation to a constant was buttressed by other dimensions of mathematical realism within mathematical discourse—by the widespread acceptance of the concept of equality within the realm of probability, for example, or the deductive form of mathematical discourse, which begins with assumptions but ends with necessity, or the generalization of correlation, which is inherently descriptive of *particular* relations. While these points are beyond the scope of this essay, they bear additional critical attention.

34. Stormer's essay comes from a special issue of *Review of Communication* that is dedicated to thinking through the implications for rhetorical theory and criticism of Karen Barad's book *Meeting the Universe Halfway* (2007). This collection of essays is an excellent starting place for those interested in exploring new materialist theories within the context of rhetorical studies.

35. As Clay Spinuzzi notes, "symmetry is a deceptively simple move: it is the principle that human and nonhuman actants are treated alike when considering how controversies are settled. . . . Its *intent* is to disrupt the assumption that only human beings have agency, acting on an inert world" (Spinuzzi 2015, 26–27).

References

Amatriain, Xavier. 2012. "Netflix Recommendations: Beyond the 5 Stars." Netflix Tech Blog, April 6. http://techblog.netflix.com/2012/04/netflix-recommendations-beyond-5-stars.html.

Aristotle. 1981. *Rhetoric*. In *Rhetoric and On Poetics*. Translated by W. R. Roberts. Edited by W. D. Ross. Franklin Center, PA: Franklin Library.

Aune, James Arnt. 2001. *Selling the Free Market: The Rhetoric of Economic Correctness.* New York: Guilford Press.

Barad, Karen Michelle. 2007. *Meeting the Universe Halfway: Quantum Physics and the Entanglement of Matter and Meaning.* Durham: Duke University Press.

Benmelech, Efraim, and Jennifer Dlugosz. 2009. "The Credit Rating Crisis." *NBER Macroeconomics Annual 2009* 24:161–207.

Burgess, Neil, Eleanor A. Maguire, and John O'Keefe. 2002. "The Human Hippocampus and Spatial and Episodic Memory." *Neuron* 35 (4): 625–41.

Charland, Maurice. 1987. "Constitutive Rhetoric: The Case of the *Peuple Quebecois.*" *Quarterly Journal of Speech* 73:133–50.

Cifoletti, Giovanna. 2006. "Mathematics and Rhetoric: Introduction." *Early Science and Medicine* 11:369–89.

Coval, Joshua, Jakub Jurek, and Erik Stafford. 2009. "The Economics of Structured Finance." *Journal of Economic Perspectives* 23:3–25.

Davis, Philip, and Reuben Hersh. 1987. "Rhetoric and Mathematics." In Nelson, Megill, and McCloskey, *Rhetoric of the Human Sciences*, 53–68.

Fabozzi, Frank J., Anand K. Bhattacharya, and William S. Berliner. 2011. *Mortgage-Backed Securities Products, Structuring, and Analytical Techniques.* 2nd ed. New York: Wiley.

Financial Crisis Inquiry Commission. 2011. *The Financial Crisis Inquiry Report: Final Report of the National Commission on the Causes of the Financial and Economic Crisis in the United States.* Official government ed. Washington, DC: Financial Crisis Inquiry Commission.

Goodnight, G. Thomas, and Sandy Green. 2010. "Rhetoric, Risk, and Markets: The Dot-Com Bubble." *Quarterly Journal of Speech* 96:115–40.

Hanan, Joshua S. 2010. "Home Is Where the Capital Is: The Culture of Real Estate in an Era of Control Societies." *Communication and Critical/Cultural Studies* 7:176–201.

Hanan, Joshua S., Indradeep Ghosh, and Kaleb W. Brooks. 2014. "Banking on the Present: The Ontological Rhetoric of Neo-Classical Economics and Its Relation to the 2008 Financial Crisis." *Quarterly Journal of Speech* 100:139–62.

Hanan, Joshua S., and Mark Hayward, eds. 2014. *Communication and the Economy: History, Value and Agency.* New York: Peter Lang.

Hariman, Robert. 1995. *Political Style: The Artistry of Power.* Chicago: University of Chicago Press.

Haug, Espen Gaarder, and Nassim Nicholas Taleb. 2011. "Option Traders Use (Very) Sophisticated Heuristics, Never the Black-Scholes-Merton Formula." *Journal of Economic Behavior and Organization* 77. http://ssrn.com/abstract=1012075.

Hick, J. R. 1948. *The Theory of Wages.* New York: P. Smith.

Houck, Davis W. 2001. *Rhetoric as Currency: Hoover, Roosevelt, and the Great Depression.* College Station: Texas A&M University Press.

International Monetary Fund. 2017. "GDP by Country, 1980–2021." International Monetary Fund Research, July 14. https://knoema.com/tbocwag/gdp-by-country-statistics -from-imf-1980-2021?country=India.

Jack, Jordynn. 2009. *Science on the Home Front: American Women Scientists in World War II.* Urbana: University of Illinois Press.

Jones, Sam. 2009. "The Formula That Felled Wall St." *Financial Times*, April 24.

Kahneman, Daniel. 2011. *Thinking Fast and Slow.* New York: Farrar, Straus and Giroux.

Krugman, Paul. 2009. *The Return of Depression Economics and the Crisis of 2008.* New York: W. W. Norton.

———. 2013. *End This Depression Now.* New York: W. W. Norton.

Latour, Bruno. 1987. *Science in Action: How to Follow Scientists and Engineers Through Society.* Cambridge: Harvard University Press.

———. 1988. *The Pasteurization of France.* Cambridge, MA: Harvard University Press.

———. 1993. *We Have Never Been Modern.* Cambridge, MA: Harvard University Press.

———. 1999. *Pandora's Hope: Essays on the Reality of Science Studies.* Cambridge, MA: Harvard University Press.

———. 2005. *Reassembling the Social: An Introduction to Actor-Network-Theory.* Oxford: Oxford University Press.

Lewis, Michael. 2010. *The Big Short: Inside the Doomsday Machine.* New York: W. W. Norton.

Li, David X. 2000. "On Default Correlation: A Copula Function Approach." http://papers .ssrn.com/sol3/papers.cfm?abstract_id=187289. Originally published in *Journal of Fixed Income* 9 (2000): 43–54.

Lynch, Paul, and Nathaniel Rivers, eds. 2015. *Thinking with Bruno Latour in Rhetoric and Composition.* Carbondale: Southern Illinois University Press.

MacKenzie, Donald. 2006. *An Engine, Not a Camera: How Financial Models Shape Markets.* Cambridge, MA: MIT Press.

Madrigal, Alexis C. 2014. "How Netflix Reverse Engineered Hollywood." *Atlantic*, January 2. http://theatlantic.com/technology/archive/2014/01/how-netflix-reverse-engineered -hollywood/282679.

Maki, Uskali. 1988. "How to Combine Rhetoric and Realism in the Methodology of Economics." *Economics and Philosophy* 4:89–109.

McCloskey, Deirdre. 1998. *The Rhetoric of Economics*. 2nd ed. Madison: University of Wisconsin Press.

McCloskey, Donald N. 1988. "Two Replies and a Dialogue on the Rhetoric of Economics: Maki, Rappaport, and Rosenberg." *Economics and Philosophy* 4:150–66.

McLean, Bethany, and Joseph Nocera. 2011. *All the Devils Are Here: The Hidden History of the Financial Crisis*. New York: Portfolio/Penguin.

Miller, Carolyn R. 2003. "The Presumption of Expertise: The Role of Ethos in Risk Analysis." *Configurations* 11:163–202.

Mirowski, Philip. 1988. "Shall I Compare Thee to a Minkowski-Ricardo-Leontief-Metzler Matrix of the Mosak-Hicks Type? Or, Rhetoric, Mathematics, and the Nature of Neoclassical Economic Theory." In *The Consequences of Economic Rhetoric*, edited by A. Klamer, D. McCloskey, and R. Solow, 117–45. Cambridge: Cambridge University Press.

———. 1991. *More Heat Than Light: Economics as Social Physics, Physics as Nature's Economics*. Cambridge: Cambridge University Press.

Morgenson, Gretchen, and Joshua Rosner. 2011. *Reckless Endangerment: How Outsized Ambition, Greed, and Corruption Led to Economic Armageddon*. New York: Times Books / Henry Holt.

Mudry, Jessica J. 2009. *Measured Meals: Nutrition in America*. Albany: SUNY Press.

Nelson, John S., Allan Megill, and Deirdre N. McCloskey, eds. 1987. *The Rhetoric of the Human Sciences: Language and Argument in Scholarship and Public Affairs*. Madison: University of Wisconsin Press.

Nussbaum, Martha. 2001. *The Fragility of Goodness: Luck and Ethics in Greek Tragedy and Philosophy*. Cambridge: Cambridge University Press.

Plutarch. 1968. *Life of Marcellus*. In *Plutarch's Lives*, vol. 5, translated by Bernadotte Perrin. Loeb Classical Library 87. Cambridge, MA: Harvard University Press.

Prenosil, Joshua. 2015. "Bruno Latour Is a Rhetorician of Inartistic Proofs." In Lynch and Rivers, *Thinking with Bruno Latour*, 97–114.

Reyes, G. Mitchell. 2004. "The Rhetoric in Mathematics: Newton, Leibniz, Their Calculus, and the Rhetoric of the Infinitesimal." *Quarterly Journal of Speech* 90:163–88.

———. 2006. "The Swift Boat Veterans for Truth, the Politics of Realism, and the Manipulation of Vietnam Remembrance in the 2004 Presidential Election." *Rhetoric and Public Affairs* 9:571–600.

———. 2014. "Stranger Relations: The Case for Rebuilding Commonplaces Between Rhetoric and Mathematics." *Rhetoric Society Quarterly* 44:470–91.

———. 2019. "Algorithms and Rhetorical Inquiry: The Case of the 2008 Financial Collapse." *Rhetoric and Public Affairs* 22 (4): 569–614.

Rotman, Brian. 1993. *Ad Infinitum: The Ghost in Turing's Machine*. Stanford: Stanford University Press.

———. 1987. *Signifying Nothing: The Semiotics of Zero*. London: Macmillan.

———. 2000. *Mathematics as Sign: Writing, Imagining, Counting*. Stanford: Stanford University Press.

Russell, Bertrand. 1907. "The Study of Mathematics." *New Quarterly* 1:29–44.

Salmon, Felix. 2012. "The Formula That Killed Wall Street." *Significance* (February 8): 16. Originally published in *Wired*, 2009.

Simonson, Peter. 2014. "Reinventing Invention, Again." *Rhetoric Society Quarterly* 44:299–322.

Sparrow, Betsy, Jenny Liu, and Daniel M. Wegner. 2011. "Google Effects on Memory: Cognitive Consequences of Having Information at Our Fingertips." *Science* 333:776–78.

Spinuzzi, Clay. 2015. "Symmetry as a Methodological Move." In Lynch and Rivers, *Thinking with Bruno Latour*, 23–39.

Stlouisfed.org. Federal Reserve Bank of Saint Louis. 2015. "Delinquency Rate on Single-Family Residential Mortgages, Booked in Domestic Offices, All Commercial Banks." Federal Reserve Bank of St. Louis Economic Research. http://research.stlouisfed.org/fred2/series/DRSFRMACBS.

Stormer, Nathan. 2004. "Articulation: A Working Paper on Rhetoric and Taxis." *Quarterly Journal of Speech* 90:257–84.

———. 2016. "Rhetoric's Diverse Materiality: Polythetic Ontology and Genealogy." *Review of Communication* 16:299–316.

Taleb, Nassim. 2007. *The Black Swan: The Impact of the Highly Improbable*. New York: Random House.

Tversky, Amos, and Daniel Kahneman. 2011 [1974]. "Judgment Under Uncertainty: Heuristics and Biases." In *Thinking Fast and Slow*, by Daniel Kahneman, 419–32. New York: Farrar, Straus and Giroux. Originally published in *Science* 185.

Wynn, James. 2012. *Evolution by Numbers: The Origins of Mathematical Argument in Biology*. Anderson, SC: Parlor Press.

5

The Ourang-Outang in the Rue Morgue | Charles Peirce, Edgar Allan Poe, and the Rhetoric of Diagrams in Detective Fiction

Andrew C. Jones and Nathan Crick

The poetical interest of the mental creation is in the creation itself, although as a part of this a mathematical interest may enter to a slight extent. Detective stories and the like have an unmistakable mathematical element. But a hypothesis, in so far as it is mathematical, is mere matter for deductive reasoning.
—Charles Sanders Peirce

"I know not," continued Dupin, "what impression I may have made, so far, upon your own understanding; but I do not hesitate to say that legitimate deductions . . . [are] sufficient to engender a suspicion which should give direction to all farther progress in the investigation of the mystery. . . . The deductions are the *sole* proper ones, and . . . the suspicion arises *inevitably* from them as the single result."
—Edgar Allan Poe

Charles Sanders Peirce had a fascination with Poe's work stretching all the way back to high school, when he performed a dramatic reading of "The Raven" (Brent 1993, 58). Explicit reference to Poe's detective fiction appears, in fact, in his 1908 essay "The Neglected Argument for the Reality of God." Peirce refers to Poe's tales of ratiocination here to defend the scientific principle that human reason possesses the capability to comprehend causes that at first seem impossible, such that "it is reasonable to assume, in regard to any given problem, that it would get rightly solved by man, if a sufficiency of time and attention were devoted to it" (Peirce 1998, 437). The reference to Poe is then used rhetorically to clinch his argument: "Those problems that at first blush appear utterly insoluble receive, in that very circumstance—as Edgar Poe remarked in his 'The

Murders in the Rue Morgue'—their smoothly-fitting keys" (1998, 437). Here, Peirce references a statement by Poe's detective, Monsieur Dupin, who says, "It appears to me that this mystery is considered insoluble for the very reason which should cause it to be regarded as easy of solution. I mean for the *outré* character of its features" (1998, 547). For Dupin, as for Peirce, what is required of an investigator of forensic, or mathematical, truth is a sustained inquiry into observable phenomena (no matter how *outré*—strange, outlandish, or bizarre— its features) that could generate a testable hypothesis based on necessary consequences. For Peirce, then, Poe's detective story was remarkable precisely for its celebration of abduction—a method that underlies much of the argumentation in general mathematical research.

Since at least the time of Euclid, mathematical arguments have been presented as nothing less than the rigorous and rigid application of deductive reasoning to even the simplest of problems—from Euclid's exhaustive proof of the Pythagorean theorem in the *Elements*, to Russell and Whitehead's several-hundred-page-effort in the *Principia Mathematica* to prove that one plus one equals two. However, as Gödel's response to Russell and Whitehead suggests, mathematical arguments can never be complete, despite, or perhaps because of, their rigorous deductive reasoning. Even before Gödel, Charles Sanders Peirce, dissatisfied with the rigidity of Euclid's approach to mathematical proof, proposed an alternative to deduction and induction that he called "abduction." Where deduction moves from the general to the specific (from a rule to a case to a result) and induction moves from the specific to the general (from a case to a result to a rule), abduction is more of a middle-out process (from a rule to a result to a case) (Bybee 1991, 286). As Brian Domino explains, abduction "attempts to show that two entities (A and B) are, at least in some sense, coextensive because they share many predicates" (1994, 63). While the modal claims of abduction are not as strong as the claims of deduction, which require the truth of a conclusion derived from true premises, or induction, which allows a high degree of certainty in the truth of a conclusion based on repeated observations, abduction does allow us to reach a degree of certainty through the addition of predicates (Bybee 1991, 286) or by identifying a predicate that satisfies all interested parties (Domino 1994, 64). For Peirce, abduction was central to the analytical reasoning through signs that he called "mathematical reasoning," because it better described how researchers came to see their solutions to a problem in a "flash of insight" like an "aha" or "Eureka!" moment (Moriarty 1996, 184).

The ideal illustration of abduction in mathematical reasoning is Edgar Allan Poe's analytical detective. If we interrogate more closely this relationship between mathematics and detective fiction with the help of Poe's story, we can better understand the role of rhetoric within mathematics—at least as it was understood by Peirce. For him, rhetoric was neither the study of necessary conclusions (which was mathematics), nor the study of reasoning through signs (which was logic), nor the art of creating beautiful images and forms (which was poetry). Rhetoric was instead the study of "the formal conditions of the force of symbols, or their power of appealing to a mind" (Peirce 1984, 57). In other words, rhetoric in mathematics is the art of making symbols effective by adapting them to the psychology of an audience in anticipation of how that audience will translate specific signs into interpretants that will then influence subsequent beliefs and actions. For instance, by referring to "The Murders in the Rue Morgue" to defend his argument that the reality of God is something that can be justified through rational argumentation, Peirce relies on his audience's knowledge of and appreciation for Poe's literary talents, thus departing from the strict chain of reasoning not only to arouse his readers' attention and interest but also to create in their minds an image that is clear and vivid and forceful. Similarly, Poe's character, Dupin, often uses necessary reasoning through signs in the context of a rhetorical situation in which he must influence others and guide events to a desirable conclusion. Peirce refers to this as a "poetic" interest, but in so far as poetry mirrors reality, we can see in the stories actual rhetorical practice in the context of forensic inquiry and mathematical reasoning.

In order to explore the relationship that mathematics has to both poetics and rhetoric, we must first elucidate how Peirce understood these relationships in theory and then explore how Poe's detective stories, which he labels tales of ratiocination, show these relationships in literary practice. These stories show the relationship between mathematics and rhetoric in two ways. First, from a poetic perspective, they demonstrate how Dupin uses a combination of logic and mathematics to solve puzzles by creating diagrams of possible relations between events to persuade other characters in the narrative to believe improbable solutions, such as that the true culprit in a murder is an escaped Ourang-Outang, or that the best way to hide something is to place it in plain sight. Second, these stories show how Poe uses the short story as a rhetorical means to advance his own methods of forensic inquiry and argumentation, in which abduction and mathematical reasoning become the primary means of assigning guilt and innocence. That is, Poe is actually trying to persuade the audience that they should use Dupin's approach to forensic inquiry and argument in every aspect of their lives. Based on this analysis, we will suggest two conclusions. First, we will argue that analytical reasoning through diagrams, which Peirce refers to as "mathematical reasoning," plays an essential role in the act of invention and should be consciously cultivated within any rhetorical art. Second, however, we will suggest that the imaginative powers that fuel analytical reasoning can lead the rhetor and the audience into error if not tempered by an attitude of fallibilism.

Peirce on Mathematics

As will become clear in our rhetorical analysis of "The Murders in the Rue Morgue," Peirce did not find the difference between mathematics and other arts and sciences where we might expect it—in the different relationship each has with numbers. As Peirce observes, there is a long tradition of thinking, ingrained deeply in both popular and intellectual culture, which associates mathematics with "the science of quantity" (2010, 16). Peirce traces this belief to Roman schoolmasters who justified this association because "the objects of the four mathematical sciences recognized by them, viz: arithmetic, geometry, astronomy, and music, are things possessing quantity" (2010, 5). What they failed to see, Peirce explains, is that "the objects of grammar, logic, and rhetoric equally possess quantity, although this ought to have been obvious even to them" (2010, 5). For instance, that rhetoric deals with quantity was particularly stressed by Aristotle in his Topics. In *The New Rhetoric* Perelman and Olbrechts-Tyteca summarize Aristotle's acknowledgment of the persuasive function of "quantitative loci" represented by arguments such as "a greater number of good things is more desirable than a smaller; a good thing useful for a comparatively large number of ends is more desirable than one useful to a lesser degree; that which is more lasting or durable is more desirable than that which is less so" (Perelman and Olbrechts-Tyteca 1971, 85–86). Because number and quantity play an important role in rhetoric, quantity cannot, therefore, be the distinguishing character of mathematics.

Following both his father and Immanuel Kant, Peirce argued that what is essential to mathematics is better understood in terms of the study of *relations* rather than of *quantity*. What really distinguishes mathematics from other disciplines "is not the subject which it treats, but its method, which consists in studying constructions, or diagrams" (2010, 17). Peirce's broadest definition of a diagram is "a mental formula always more or less general" (1998, 246). As Buchler sums up Peirce's conception, "diagrams are our hypothetical expressions arranged in definite orders by rules" (1939, 213). Peirce gives the following examples of mathematical diagrams: "The diagrams in which the [mathematician's] hypotheses are embodied are of two kinds. In the one kind [the inherential kind] the parts of the diagram are seen in the visual image to have the relations supposed. In the other kind of diagrams, the parts have shapes to which conventions or 'rules' are attached, by means of which the supposed relations are attributed, or imputed, to the parts of the diagrams. Geometrical figures are diagrams of the inherential kind, while algebraic formulae are diagrams of the imputations kind" (2010, 46–47).

In this quote Peirce discusses the difference between "inherent" (or iconic) and "imputed" diagrams, like the difference between a picture of a right triangle and the formula $a^2 + b^2 = c^2$. However, Peirce is thinking not about diagrams in Euclidean space here but about diagrams in Lobachevskian or Riemannian space. As Kragh (2012, 153) explains, Peirce was a committed non-Euclidean geometer and supported his claims with both philosophical and observational arguments. Within the context of Peirce's non-Euclidean conviction, the iconic diagram employs three lines to visually represent a relationship between the lines that is usually comprehensible on its own, whereas the imputed diagram of the formula can only be interpreted as a mathematical relationship through a system of conventions or rules. Peirce rejected Euclidean axioms that would identify the three lines of a diagram as the description of a shape. Instead, the Euclidean axioms would fit under the category of imputed diagrams, which work well enough in infinite space but cannot explain the relationships Peirce observed in his calculation of the approximate value of the curvature of space (Dipert 1977, 404). In both cases, however, the diagrams are mathematical because they represent a necessary system of relationships that can be examined, manipulated, or rearranged in the imagination. For Peirce, mathematics is not contingent on the presence of any specific subject matter or even any set of symbols. Algebraic equations, geometrical figures, numerical formulas, or singular qualities are all possible material for analytical reasoning in so far as they can be placed within a system of relations, "but the character of the objects, apart from the relations, is utterly immaterial" (2010, 46). What matters in mathematics is that some diagram exists that can be experimented on mentally to derive some conclusion about the relationships represented by the diagram.

Two characteristics distinguish a mathematical diagram from a logical, ethical, political, or rhetorical diagram. First, a mathematical diagram is utterly unconcerned with the relationship of its premises or conclusions to actual phenomena, experience, or empirical fact. Peirce explains, "*mathematical reasoning may be defined as a reasoning in which the following of the conclusion does not depend on whether the premises represent experience, or represent the state of the real universe, or upon what universe it may be that they apply to*" (2010, 92). Second, a mathematical diagram is used to study the relations inherent in the diagram. While this may seem tautological at first, the interest is not in the use of a diagram to study a diagram but in the use of a diagram to study the relationships that become clearer by studying their character, transformations, and relations to one another. A mathematician thus "observes nothing but the diagrams he himself constructs; and no occult compulsion governs his hypothesis except one from the depths of mind itself" (2010, 4). By contrast, an accountant who also works with numbers all day would be a logician rather than a mathematician because his or her interest is in what the numbers *actually refer to* in the objective world, while a mathematician would only be concerned with the possible calculations across a spreadsheet.

Our interest is in the iconic diagram more than the imputed diagram of the algebraic formula because iconic diagrams illustrate the relationships between available data through images; whereas imputed diagrams emphasize the rules and conventions governing the symbols as expressed through formulas. Iconic diagrams still require conventions to identify that aspects of the diagram are relevant, but because they generally resemble what they represent, they are able to convey more information than imputed diagrams and, therefore, they have more persuasive force within an argument based on abduction. As Bybee explains, "multiplying the number of conditions an abductive conclusion must satisfy constricts the number of conclusions capable of satisfying those conditions" (1991, 286). Where imputed diagrams have an aura of certainty about them because they are based on deduction and, therefore, their conclusions must be true so long as all of their premises are true, iconic diagrams are more important for rhetoric in mathematical reasoning because researchers aim not at certainty but at belief in their first presentation of a solution.

As Hookway (2007) notes, the representative aspect of an iconic diagram is like a map. In order to use it, we must know the conventions governing its symbols: it must be anchored, and it must be analogous. The resemblance between an iconic diagram and the relationships it represents can be very abstract, like a map drawn on the back of a napkin, or detailed, like a Google satellite map. What is important is how the iconic diagram can be used. A better metaphor for our investigation of analytic detective fiction might be the murder board investigators use to represent different relationships within a case. An iconic diagram is anchored by points of reference, as when a murder board includes marks that represent key evidence in the case. An iconic diagram is analogous to the thing it represents when we can imaginatively manipulate the elements of the diagram in order to plan our actions, as when a detective moves a suspect's photograph on a murder board to consider how the case would look if the person were actually the intended victim. Because it is anchored and analogous to the thing it represents, "we can obtain information about the terrain [the investigation] by making observations of the sign [the murder board]: we exploit the known similarity in order to infer from properties of the sign to properties of the object" (Hookway 2007, 61). Furthermore, because the iconic diagram is analogous to the thing it represents, "we can make experiments upon the map to extend our knowledge of the terrain" (2007, 61). For example, detectives can hypothesize the effects of a missing piece of evidence by making marks on a murder board and observing how this will affect relationships between the victim, the suspects, the evidence, and the scene, all of which allow detectives to navigate those relationships and test their hypotheses.

Most of our everyday hypotheses, of course, require no conscious or complex mathematical reasoning with diagrams. However, this type of reasoning is called for when an individual is "confronted with an unusually complicated state of relations between facts and is in doubt whether or not this state of things necessarily involves a certain other relation between facts, or wishes to know what relation of a given kind is involved" (Peirce 2010, 3). The task of the mathematician is thus "to substitute for the intricate, and often confused, mass of facts set before him, an imaginary state of things involving a comparatively orderly system of relations, which, while adhering as closely as possible or desirable to the given premises, shall be within his powers as a mathematician to deal with" (2010, 3). The mathematician *qua* mathematician, of course, remains fundamentally unconcerned with the empirical validity of any possible conclusions. His or her concern is simply to "show that the relations explicitly affirmed in the hypothesis involve, as a part of any imaginary state of things in which they are embodied, certain other relations not explicitly stated" (2010, 3). In the case of someone like a police detective, however, it makes all the difference in the world what relations are inferred and whether those other relations are *merely possible* or actually exist in the present case.

Rhetoric *in* Mathematics

It is notable that one of Peirce's definitions of a diagram as a mental picture of active relationships appears in his discussion of ethical conduct in which an individual is struggling to make a moral judgment. He describes, for instance, how all individuals carry with them certain conceptions of ideal conduct to which they imagine themselves committed. Often, however, these ideals remain largely abstract until one is confronted with an actual situation. During these times, particularly when confronted with situations in which alternative actions are possible, an individual "imagines what the consequences of fully carrying out his ideals would be, and asks himself what the aesthetic quality of those consequences would be" (Peirce 1998, 246). For Peirce, aesthetics is the study of "objects considered simply in their presentation" (1998, 143). Aesthetics represents the capacity to identify and appreciate the qualities associated with a certain act, event, person, or object, whether those qualities are pleasurable or distasteful. This aesthetic visualization then provides material with which the moral imagination can work in order to "consider how he will act"—namely by identifying and tracing out the specific consequences that one can anticipate from such action (1998, 246). The consequence of bringing together these rules of conduct with their anticipated aesthetic consequences within a particular situation then produces effects on the will in the form of "a *resolution* as to how he will act upon that occasion" (1998, 246). Importantly, Peirce adds the following: "This resolution is of the nature of a plan, or, as one might say, a *diagram*" (1998, 246). Diagrammatic thinking is thus clearly essential not only to mathematical reasoning but also to logical, ethical, and rhetorical reasoning as well.

But what does Peirce mean, exactly, by rhetoric? For John Lyne, Peirce understood rhetoric as "the art of making signs effective" and the study of the "ways signs are put to use" (1980, 165). For Vincent Colapietro, the "rhetorical question in the Peircean sense concerns, in any usage of signs over which self-control is in some measure possible, how to render signs efficacious or effective and also fruitful or fecund" (2007, 27). In his own writing, Peirce defined the formal study of rhetoric as "the science of the essential conditions under which a sign may determine an interpretant sign of itself and of whatever it signifies, or may, as a sign, bring about a physical result" (1998, 482). In other words, for Peirce rhetoric is the study of how any signs—whether icons, indices, or symbols—are communicated in such a way that they appeal to a specific mind or minds, namely by anticipating not only the interpretants that any particular audience will associate with those signs but also the subsequent beliefs, emotions, habits, or actions that result from them. He offers the following broad account of this definition:

> [Let us acknowledge] a universal art of rhetoric, which shall be the general secret of rendering signs effective, including under the term "sign" every picture, diagram, natural cry, pointing finger, wink, knot in one's handkerchief, memory, dream, fancy, concept, indication, token, symptom, letter, numeral, word, sentence, chapter, book, library, and in short whatever, be it in the physical universe, be it in the world of thought, that, whether embodying an idea of any kind (and permit us throughout to use this term to cover purposes and feelings), or being connected with some existing object, or referring to future events through a general rule, causes something else, its interpreting sign, to be determined to a corresponding relation to the same idea, existing thing, or law. (1998, 326)

Notably—as indicated by his inclusion of numerals and diagrams—Peirce in his definition rejects any strict dichotomy between rhetoric and science. For him, rhetoric includes *any* and *all* ways of making signs effective, including those "rules of expression" that regulate the most technical sciences (1998, 326). As he observes, "a proposition of geometry, a definition of a botanical species, a description of a crystal or of a telescopic nebula is subjected to a mandatory form of statement that is artificial in the extreme. What is the principal virtue ascribed to algebraic notation, if it be not the rhetorical virtue of perspicuity?" (1998, 326). Consequently, Peirce recommends removing the "restriction of rhetoric to speech" and of generalizing rhetoric to include all uses of signs—including strictly mathematical signs—to arouse specific interpretants in other minds that have predictable effects (1998, 326).

But, as we have been describing, another way to look at rhetoric—and to connect it productively with mathematics—is to the language of a diagram. In this definition, rhetoric uses signs to suggest to another mind a certain diagram that is, as Peirce mentioned in the context of conduct and judgment, "of the nature of the plan" (246). In this diagram, one might outline a system of relations that incorporates what are taken to be the established facts of any situation, the rules of conduct that apply to that situation, and the effects that might be or have been produced from a specific course of action. In his lengthy description of what accounts for a rhetorical sign, for instance, Peirce specifies what amounts to three different types of signs that might be incorporated with any diagram. These include those that embody "an idea of any kind" (that is, an Icon of a Firstness), those that are "connected with some existing object" (an Index of a Secondness), or those that refer to "future events through a general rule" (a Symbol of a Thirdness) (1998, 326). Rhetoric, in other words, constructs persuasive diagrams that vividly bring before the eyes certain salient qualities of any situation, connects these qualities with specific objects that bear upon judgment, and provides the rules, principles, norms, and laws that specific audiences can use to regulate future events and attain definitive purposes.

Although rhetorical scenarios deal with real objects, contingent affairs, real audiences, and practical judgments in a way that mathematical diagrams do not, logical argumentation—which has ties to both mathematics and rhetoric—clearly plays a role within rhetorical invention. The act of invention, that is to say, traces back both to Aristotle's definition of rhetoric as "an ability, in each [particular] case, to see the available means of persuasion" and to his definition of persuasion as that which occurs "through the arguments [*logoi*] when we show the truth or the apparent truth from whatever is persuasive in each case" (2007, 1355b, 1356a). What mathematical reasoning adds to the act of invention is the ability for an active participant in a situation to step back and take on the perspective of an analytic observer. Adopting the perspective of the mathematician allows one the ability to distance oneself from the contingency and flux of any situation, to identify persons, events, and objects, and then to create a diagram that accounts for their interaction and predicts their potential consequences based on the application of specific principles and rules. In other words, mathematical reasoning makes possible what Peirce calls acts of "insight" that come "to us like a flash" (1998, 227). As he explains, although in many cases "the different elements of the hypothesis were in our minds before," it is "the idea of putting together what we had never before dreamed of putting together which flashes the new suggestion before our contemplation" (1998, 227). As Poe's detective Monsieur Dupin so masterfully demonstrates, it is the analytical aspects of invention (of stepping back to find the available means of persuasion by the recognition of new patterns of relations) that make possible rhetorical invention.

Dupin's conclusions also show the limitations of analytical reasoning: in practical affairs it must always be kept in mind that mathematical idealizations cannot be guaranteed to track with reality. Mathematical insight, that is to say, is often an "extremely fallible insight" (1998, 227). It is notable, for instance, that Peirce argues that "the typical Pure Mathematician is a sort of Platonist. . . . The Eternal is for him a world, a cosmos, in which the universe of actual existence is nothing but an arbitrary locus" (40). But to act in the universe of actual existence as if it were a perfectly formulated cosmos run by predictable, rational, eternal laws is to mistake contingencies for necessities and to treat particularities as if they were simply instantiations of some universal definition. As Perelman observes, although it is tempting to "reduce all deductive reasoning to a demonstration which would be correct if the operations agreed with a preestablished scheme," in rhetorical practices of argumentation, "we have seen that not only facts and truth can be questioned, but that even the determination of what the datum is, is contingent upon the result of a discussion concerning its interpretation" (1982, 48). In the context of the narrative in Poe's detective fiction, Dupin's hypotheses prove themselves uncannily accurate. This accuracy, we argue, is a rhetorical performance arguing that mathematical reasoning can be effectively applied in *everyday life*.

The Murders in the Rue Morgue

Peirce looks to Poe's detective story for an example of analytical reasoning because Poe's forensic analyst and Peirce's mathematician are one and the same. Poe's story "The Murders in the Rue Morgue," which was originally published in *Graham's Magazine*, begins with a philosophical treatise on the mind, advancing the proposition that analysis, what Peirce calls mathematical reasoning, is the combination of calculation, discrimination, and imagination. Poe then subordinates the narrative to a "commentary upon the proposition just advanced" (1978a, 531). Poe begins his treatise by opposing "the vulgar dictum (founded, however, upon the assumptions of grave authority), that the calculating and discriminating powers (causality and comparison) are at variance with the imaginative—that the three, in short, can hardly coexist" (1978a, 527). Instead, Poe argues, the process of invention (creation) is the inverse of the process of resolution (analysis) because both require the calculation of causes, comparison of differentia, and imagining of alternatives. That is, the forensic analyst solves a case through the process of resolution by schematizing the essential evidence and mapping known causes to imagine necessary relationships that are otherwise obscured, thereby resolving the case through an iconic diagram.

The entire work—frontispiece, preface, narrative, and all—is an encomium to the analyst, a rhetorical hymn in praise of those who use mathematical reasoning in their everyday life. Poe compares the analyst to an athlete engaged in a competition, arguing that "as the strong man exalts in his physical ability, delighting in such exercises as call his muscles into action, so glories the analyst in that moral activity which *disentangles*. He derives pleasure from even the most trivial occupations bringing his talents into play" (1978a, 528). As Peirce suggests, the moral dimension of resolution is that it allows the analyst to consider the consequences of certain actions, and then to act according to his convictions and his vision of the future. In addition to reveling in the moral and intellectual strength of the analyst, Poe also ascribes an almost occult character to him: "He is fond of enigmas, of conundrums, of hieroglyphics; exhibiting in his solutions of each a degree of *acumen* which appears to the ordinary apprehension praeternatural. His results, brought about by the very soul and essence of method, have, in truth, the whole air of intuition" (1978a, 528). That is, the analyst appears god-like through an ability to suss out the hidden truths of the universe and understand their relations and meanings, but this appearance is based on an underlying method.

Poe forms connections between the underlying method of abduction and mathematical reasoning, thus prefiguring Peirce's argument that mathematics is not merely the science of quantity but the study of necessary relations. Poe explains that analysis is not mere calculation, the determination of the size or number of things, but the exploration of relationships. According to Poe, "the faculty of re-solution is possibly much invigorated by mathematical study, and especially by that highest branch of it which, unjustly, and merely on account of its retrograde operations, has been called, as if *par excellence*, analysis" (1978, 528). That is, the methods of mathematical study prepare the analyst to dazzle others with their *outré* revelations.

Poe illustrates this point through an analogy comparing the games of chess and draughts (checkers). He argues that chess is similar to algebra, a game of elaborate frivolity "where pieces have different and *bizarre* motions, with various and variable values" (1978a, 528). That is, chess players must simply follow the rules of the game—plugging values into a well-known equation in order to calculate the available moves—until a player makes a miscalculation and, due to the complexity of the game, loses. Poe explicitly argues that it is attention and not analytical thinking that gives the chess player an advantage: "in nine cases out of ten it is the more concentrative rather than the more acute player who conquers" (1978a, 528). Checkers, however, is a game of analysis because "the moves are unique and have but little variation," so that one must play the player as much as the game (1978a, 528). "Deprived of ordinary resources," Poe explains, "the analyst throws himself into the spirit of his opponent, identifies himself therewith, and not unfrequently sees thus, at a glance, the sole methods (sometimes absurdly simple ones) by which he may seduce into error or hurry into miscalculation" (1978a, 529). That is, Poe—like Peirce—sees mathematics not as a simple game of chess, governed by rules and formulae for movement, but as a game of relations, wherein every possible relationship—including the relationship between the players engaged in the game—may have bearing on the game.

Poe's analyst and Peirce's mathematician are one and the same. Their ability is derived not from chance and good fortune but from the ability to acquire data from their surroundings, detach themselves from those same surroundings, and form new hypotheses to explain the relationships they observe in order to consider new possibilities of thought and action. Unlike the Roman orator considering prefabricated lists of motives or a geometer relying on the axioms laid down by Euclid, Poe's analyst and Peirce's mathematician are responsible for considering new motives, for thinking outside the infinite plane of Euclidean geometry and considering the geometry of a curved space. To return to Poe's metaphor of the game, the mathematician is able to detach from the assigned rules of the game and invoke new ones that creatively assess the widely heterogeneous possibilities presented by opponents and environments as well as the conventional pieces and moves on the board.

Poe illustrates his proposition that analysis is the combination of calculation, discrimination, and imagination through the different approaches of the police and the detective in his detective stories. The police simply plug evidence into the imputed diagram of a preexisting formula, whereas the analytic approach of the detective allows for the envisioning of an inherent diagram allowing the detective to consider multiple possible relationships and thus identify solutions to an intractable case. That is to say, when the detective enters a new room he searches for a hypothesis that best accounts for the consequences and constraints he observes, even if this forces him to abandon established case theory. Poe implies that the power of the analyst extends to the rhetorical power of invention—the ability to discover available means of persuasion. This acumen for invention is particularly important when the local authorities, the police, are baffled by a particularly *outré* crime that cannot be solved by guesswork and persistence alone. Such was the case of the murders in the Rue Morgue.

Shrieks pierce the air of Quartier St. Roch at three o'clock one morning, bringing a crowd to the locked gates of Madame L'Espanaye's four-story house in the Rue Morgue. Isidore Musèt, a gendarme, bursts through the gate with his bayonet and leads a party of neighbors up the stairs to a locked room. After breaking down the door, they find the room empty and disheveled and the fresh corpse of Mademoiselle Camille L'Espanaye head-downward in the chimney. The body of her mother, Madame L'Espanaye, lies nearly decapitated in the courtyard. A thorough search of the building is conducted, but the murderer is not found. All witnesses agree that the doors and windows to the room containing the mademoiselle's corpse were locked from the inside and no one could have passed down the stairs while the gendarme led a posse up the stairs. After taking statements from twelve witnesses, the following additional facts come to light. After the screams ceased, two voices were heard. One was of a Frenchman shouting "*sacré*," "*diable*," and "*mon Dieu*." The other voice was variously reported as both male and female, both shrill and harsh, and speaking Italian, Spanish, Russian, English, and French. In each case the language assigned was foreign and incomprehensible to the witness. Additionally, the print of a hand was discovered in the bruises on the mademoiselle's throat, and reddish hairs were found in the grasp of the madame. At a loss, the Parisian police arrest a clerk named Adolphe Le Bon, who had delivered four thousand francs to the apartment the day before, even though they have no evidence of his involvement in the murders and none of the gold, Le Bon's only presumable motive, has been removed from the room.

Enter Dupin, an analyst whose talent for picking up on signs, interpreting them, and creating hypothetical explanations and predictions makes him exceptional at abductive reasoning. Dupin reduces the problem of the locked room to an iconic diagram in his imagination. Dupin's diagram is ruled by the simple convention that "all apparent impossibilities *must* be proved to be not such in reality" (Poe 1978, 552). In order to fit the victims, suspects, and evidence into such a diagram, Dupin abstracts three key pieces of evidence from the testimony of the witnesses—which he gathers from the *Gazette des Tribunaux*. The first piece of evidence is the discrepancy in reports about the voice, the second is the means of ingress and egress, and the third is the absence of motive. These are the *outré* problems of the case that, according to both Peirce and Poe, should give the case its smoothly fitting keys. But before establishing his final argument, Dupin experiments with his conceptual diagram of the case. In doing so, he perfects his mental picture of the relationships between the data by imagining the missing links in these relationships. As he explains to his interlocutor, "it is by these deviations from the plane of the ordinary, that reason feels its way, if at all, in the search for the true. In investigations such as we are now pursuing, it should not be so much asked 'what has occurred,' as 'what has occurred that has never occurred before'" (Poe 1978a, 548). Rather than ignore the irregularities of the case and focus only on a formula, like the common means, motive, and opportunity of a police procedural, Dupin focuses on the irregularities of the case to consider new possibilities. However, Dupin—as detective—must use rhetoric to persuade others that his hypothetical diagram fits this particular crime and not merely this type of crime in an abstract universe of discourse. In order to proceed, Dupin first devises a hypothesis, then he tests his hypothesis, and finally he takes his confirmed hypothesis to the authorities and persuades them to take action.

Step One: Devising a Hypothesis

For a mathematician interested only in abstract diagramming, investigation of a murder scene would be a mere amusement because the reported facts of the case would be sufficient to arrive at a solution through abductive reasoning. However, for a forensic rhetorical analyst like Dupin, the facts of the case reported by the newspaper would not provide sufficient or necessarily compelling evidence to persuade others to accept his hypothesis about the murder. For that, it was necessary to examine the scene. Having developed an iconic diagram from the reported evidence through abduction, Dupin mentally probed the relationships among the pieces of evidence to identify missing features of the diagram—most notably, the perpetrator of the butcheries. The first piece of evidence, which confounded the police but aided Dupin, was "the seeming impossibility of reconciling the voices heard in contention, with the facts that no one was discovered upstairs but the assassinated Mademoiselle L'Espanaye, and that there were no means of egress without the notice of the party ascending" (Poe 1978a, 547). Dupin first dismisses the idea that the voices in the room could have been the voices of the two women, because the old woman could not have thrust her daughter's body feet first up the chimney, nearly decapitated herself, and jumped through a window, locking it behind her. Therefore, Dupin concludes, the voices must belong to some third party who is responsible for the murder. The point the police have missed is that, while all the witnesses agreed that the gruff voice belonged to a Frenchmen, they disagreed about the other voice. "That is the evidence itself," Dupin explains, "not that they disagreed—but that, while an Italian, and Englishman, a Spaniard, a Hollander, and a Frenchman attempted to describe it, each one spoke of it as that *of a foreigner*" (1978, 549). Though the exclamations of the gruff voice could be recorded in French, Dupin points out that the other voice was harsh, quick, and unequal, and no sounds resembling words were mentioned, even though they were supposedly heard by "denizens of the five great divisions of Europe" (1978, 550). Dupin then adds a new relationship to his iconic diagram, though he does not share it with anyone yet. He adds the possibility that the harsh voice belonged to no human but to an animal.

At this point, Dupin has a hypothetical solution to the case. Hypothesizing an animal as the culprit explains the evidence of the incomprehensible voice, the lack of motive, and the inhuman brutality of the murders. Although abductive reasoning provides a reasonable rationale for Dupin, persuading others requires the formation and testing of a hypothesis. Only with additional evidence does he believe his reasoning will persuade others that his abductions "are the *sole* proper ones, and that the suspicion arises *inevitably* from them as the single result." Thus he turns to the second key fact of his iconic diagram, the means of egress, and gains the permission of the Prefect of police to inspect the locked room for himself.

First, Dupin observes the chamber, which was not as carefully described in the papers as the witness testimony about the voices had been. Under the assumption that "the doers of the deed were material, and escaped materially," he carefully examines the entire neighborhood as well as the chambers of the fourth floor. Finding no secret passageways, no unlocked doors, and no passable chimneys, he concludes that the murderers *must* have exited through a window in the back room (Poe 1978a, 551). However, he finds that both windows in the back room are securely latched with "clews," a type of nail, apparently driven into the casement. Dupin uses abductive reasoning to transcend the apparent impossibility of a latched and secured window, by considering that, if there is no other means of entering or exiting the room, then the window *must not* be actually nailed shut. He ruminates:

The murderers did escape from one of these windows. This being so, they could not have re-fastened the sashes from the inside, as they were found fastened;—the consideration which put a stop, through its obviousness, to the scrutiny of the police in this quarter. Yet the sashes were fastened. They must, then, have the power of fastening themselves. There was no escape from this conclusion. I stepped to the unobstructed casement, withdrew the nail with some difficulty, and attempted to raise the sash. It resisted all my efforts, as I had anticipated. A concealed spring must, I now knew, exist; and this corroboration of my idea convinced me that my premises, at least, were correct, however mysterious still appeared the circumstances attending the nails. A careful search soon brought to light the hidden spring. I pressed it, and, satisfied with the discovery, forbore to upraise the sash. (Poe 1978a, 552)

Step Two: Confirming the Hypothesis

Dupin's next move is to test all of his hypotheses by luring a witness into testifying who, "although perhaps not the perpetrator of these butcheries, must have been in some measure implicated in their perpetration" (Poe 1978a, 548). Dupin offers reasoning and supplemental evidence to his unnamed interlocutor as they wait with loaded pistols for the man responsible for the crime (1978a, 548). Dupin's explanations to his sidekick, the unnamed narrator of Dupin's adventures, allow Poe to demonstrate to the reader how the case was solved through Dupin's analytical powers of abductive reasoning while presenting additional evidence to support Dupin's hypotheses. By schematizing the necessary relationships into an iconic diagram that is anchored to the case by key evidence, the detective invents arguments that better explain the relationships of the diagram and thus solves the case. That is, the physical clues guide Dupin's reasoning but are not themselves necessary for reaching his conclusion. While the voices persuaded Dupin that an ape was the only possible solution to the mystery, it was the apparent "clew" in the window casement that persuaded the police that the mystery of the Rue Morgue was unsolvable. When Dupin reasoned that the window was the only possible means of egress, and that it must possess a self-latching mechanism, he knew that the "clew" in the casement must be false. The evidence was less important to the solution of the case than the iconic diagram in Dupin's mind, and, in the case of the "clew," the evidence was actually an impediment to finding a reasonable solution to the case.

By the time the entrapped witness of the crime arrives, Dupin has already explained to the narrator how the description of the voices at the crime scene led him to the solution of the case, and how the physical evidence from the murder scene led him to the identity of the crime's witness and perpetrator. However, he withholds the identity of the perpetrator until he has led his interlocutor through the details of the case. He introduces the problem of the voices but does not reveal the owner of the harsh voice until he has illustrated three additional attributes of the perpetrator. First, that the perpetrator had to possess uncommon agility to climb up a lightning rod, leap to a shutter, and swing himself into the room. Furthermore, the brutality of the crime—slicing through the flesh, sinew, and bone of the older woman's neck—and the superhuman strength required to carry it out—wedging the younger woman up the chimney so thoroughly that "the united vigor of several persons was found barely sufficient to drag it *down!*"—suggested a nonhuman agent (Poe 1978a, 557). Second, the perpetrator left the old woman's money, the only conceivable motive for the murders, behind at the scene. The police arrested Le Bon because he had delivered four thousand francs in gold to the apartment three days before the murders, but "under the real circumstances of the case, if we are to suppose gold the motive of this outrage, we must also imagine the perpetrator so vacillating an idiot as to have abandoned his gold and his motive together" (1978a, 556). After connecting this evidence, Dupin pauses and allows the narrator to guess at the solution: "A madman," he suggests, but Dupin offers two pieces of physical evidence to challenge that possible hypothesis. The third attribute of the perpetrator is a tawny tuft of hair, clutched in the rigid fingers of Madame L'Espanaye, and the marks left on the throat of Mademoiselle L'Espanaye. The hair is clearly not human hair, and when Dupin presents a tracing of the marks from the mademoiselle's throat, his partner, unable to place his fingers in the same positions, exclaims, "This . . . is the mark of no human hand" (1978a, 559). Having led his companion and the reader to a juxtaposition of these three attributes, Dupin gives his partner a passage from Cuvier's description of the "large fulvous Ourang-Outang of the East Indian Islands" (Poe 1978a, 559).

As Dupin concludes his proofs, a step is heard on the stairs. Dupin has lured the Ourang-Outang's negligent owner with the following advertisement in the newspaper: "CAUGHT—*In the Bois de Boulogne, early in the morning of the—inst.,* (the morning of the murder,) *a very large, tawny Ourang-Outang of the Bornese species. The owner, (who is ascertained to be a sailor, belonging to a Maltese vessel,) may have the animal again, upon identifying it satisfactorily, and paying a few charges arising from its capture and keeping. Call at No.—, Rue—,' Faubourg St. Germain—au troisième"* (1978a, 560–61). Dupin admits that the man's identity is not as certain as the animal's: "I am not *sure* of it. Here, however, is a small piece of ribbon, which from its form, and from its greasy appearance, has evidently been used in tying the hair in one of those long *queues* of which sailors are so fond. Moreover, this knot is one which few besides sailors can tie, and is peculiar to the Maltese" (1978a, 561). Dupin guesses that, even if he is wrong about the particulars of the man's profession, the advertisement is sufficient to draw out the animal's owner because there is no apparent connection to the crime in the Rue Morgue—the advertisement states that the animal was caught a great distance from the Rue Morgue, and the police "have failed to procure the slightest clew" in their investigations of the murder (1978a, 561). Therefore, Dupin opens the door and welcomes a sailor into his apartment.

Step Three: Persuading the Authorities

Once the sailor is trapped with Dupin and the narrator, Dupin manages the situation so that there is no possibility of the sailor denying his responsibility for the murders. Like a player of draughts, Dupin traps him by making it seem as though he can predict all potential moves. The fact that all of the evidence in the case has already been determined allows him to make legal and moral arguments. First, he invites the sailor to be forthright in his testimony by explaining, "You were not even guilty of robbery, when you might have robbed with impunity. You have nothing to conceal. You have no reason for concealment" (Poe 1978a, 564). However, by mentioning the possibility of robbery, Dupin shows that he knows the sailor was in the murder room and saw the crime. The sailor breaks down under the weight of Dupin's arguments and confesses everything: how a shipmate died, leaving behind the beast, how the Ourang-Outang had escaped one evening with a shaving razor, how it had climbed—exactly as Dupin hypothesized—up the lightning rod and in through the window, and finally how it had slain the occupants of the house on the Rue Morgue. The sailor offers the following graphic account that confirms the details of the iconic diagram that Dupin had constructed:

As the sailor looked in, the gigantic animal had seized Madame L'Espanaye by the hair, (which was loose, as she had been combing it,) and was flourishing the razor about her face, in imitation of the motions of a barber. The daughter lay prostrate and motionless; she had swooned. The screams and struggles of the old lady (during which the hair was torn from her head) had the effect of changing the probably pacific purposes of the Ourang-Outang into those of wrath. With one determined sweep of its muscular arm, it nearly severed her head from her body. The sight of blood inflamed its anger into frenzy. Gnashing its teeth, and flashing fire from its eyes, it flew upon the body of the girl, and imbedded its fearful talons in her throat, retaining its grasp until she expired. Its wandering and wild glances fell at this moment upon the head of the bed, over which the face of its master, rigid with horror, was just discernible. The fury of the beast, who no doubt bore still in mind the dreaded whip, was instantly converted into fear. Conscious of having deserved punishment, it seemed desirous of concealing its bloody deeds, and skipped about the chamber in an agony of nervous agitation; throwing down and breaking the furniture as it moved, and dragging the bed from the bedstead. In conclusion, it seized first the corpse of the daughter, and thrust it up the chimney, as it was found; then that of the old lady, which it immediately hurled through the window headlong. (1978a, 566–67)

Unfortunately, Poe does not record Dupin's conversation with the Prefect, but he does state that Dupin accompanies the sailor to the Prefect and explains the facts in the case, securing the release of Le Bon and preventing the incarceration of the sailor.

In this story, Dupin's abductive reasoning is essential for the forensic investigation into the case. Abductive reasoning encourages Dupin to consider the otherwise absurd possibility that an Ourang-Outang has committed the murders in the Rue Morgue, where the police—bound by their formula of means, motive, and opportunity—had arrested an innocent person because he might have had a motive. Dupin's abductive reasoning also encourages him to consider the otherwise absurd possibility that the murderer fled through a window that was both latched and nailed shut. That is, the window had to be the point of egress, regardless of the "fact" of the locked window in the real universe. His test of the relationship between the window and the rest of the case is performed not only for his own satisfaction but also to later persuade his interlocutor that his apparently absurd conclusion reached through analytical reasoning was in fact correct. Likewise, Dupin persuades the sailor to confess by revealing that his abductive process had provided him with "means of information about this matter—means of which you could never have dreamed" (1978a, 564). Thus the force of Dupin's persuasion comes from a combination of his analytical reasoning, which allows him to transcend the facts of the case and see its possible solutions, and rhetoric, which he uses to encourage the sailor to confess using a narrative that makes Dupin's iconic diagram of the case appear possible despite its implausibility.

The Dangers of Abductive Reasoning: "The Mystery of Marie Rogêt"

Poe's fiction exemplified the kind of reasoning he believed needed to be applied in actual cases of homicide. In 1842 he would be given the opportunity to test his abductive skills in "The Mystery of Marie Rogêt." As Poe writes in a letter to potential publisher George Roberts, the "story is based upon the assassination of Mary Cecilia Rogers, which created so vast an excitement, some months ago, in New York. I have, however, handled my design in a manner altogether *novel* in literature" (Poe 2008, 337). What was novel about the story was that Poe used the reported facts of the case to construct a narrative of the murder to demonstrate the power of abductive reasoning in solving real-world crimes. Poe writes in a letter attempting to sell the story to Joseph Snodgrass of the *Baltimore Saturday Visitor*, "my main object, however, as you will readily understand, is the analysis of the *principles of investigation* in cases of like character. Dupin *reasons* the matter throughout" (2008, 340). And for a mere $50, Poe offers James Herron the solution to the Mary Rogers's case "under presence of showing how Dupin . . . unraveled the mystery of Marie's assassination" (2008, 338). By carefully scrutinizing the newspapers, which had been reporting almost every detail of the case since Mary Roger's disappearance in late July of 1841, Poe claimed not only to have disproved all of the major theories about the case but also to "have *indicated the assassin* in a manner which will give renewed impetus to investigation" (2008, 338). Poe's detective story would reassemble the facts reported in the newspapers to persuade the public of the method's efficacy in solving real-world crimes through abduction in mathematical reasoning.

Unlike the other Dupin stories, which focus on presenting evidence, a solution, and then an explanation of Dupin's reasoning, the majority of "The Mystery of Marie Rogêt" is made up of Dupin's refutation of current theories of the case. Citing American newspapers, renamed to fit with the imagined Parisian setting, Poe uses Dupin to argue that the body dragged from the water was that of Marie Rogêt (Mary Rogers), that she was not the victim of a gang of ruffians—as the papers argued—but of a single assassin. Dupin explains:

> Let us sum up now the meagre yet certain fruits of our long analysis. We have attained the idea either of a fatal accident under the roof of Madame Deluc, or of a murder perpetrated, in the thicket at the Barrière du Roule, by a lover, or at least by an intimate and secret associate of the deceased. This associate is of swarthy complexion. This complexion, the "hitch" in the bandage, and the "sailor's knot," with which the bonnet-ribbon is tied, point to a seaman. His companionship with the deceased, a gay, but not an abject young girl, designates him as above the grade of the common sailor. Here the well written and urgent communications to the journals are much in the way of corroboration. The circumstance of the first elopement, as mentioned by Le Mercurie, tends to blend the idea of this seaman with that of the "naval officer" who is first known to have led the unfortunate into crime. (1978b, 768–69)

While Poe's story did coincide with renewed interest in the investigation, he did not succeed in indicating the assassin, primarily because new evidence came to light between the second and third installments of his story which disproved his theorem. Poe attempted to fix the story by holding back the final installment and altered the ending to explain that what worked in Paris might not work in New York because "the most trifling variation in the facts of the two cases might give rise to the most important miscalculations" (1978b, 773). Poe then compares the story to a mathematical problem, writing that "very much as, in arithmetic, an error which, in its own individuality, may be inappreciable, produces, at length, by dint of multiplication at all points of the process, a result enormously at variance with truth" (1978b, 773). Literally, the comparison suggests that both the detective and the mathematician can fall prey to compounded error, but figuratively the comparison highlights Poe's fundamental error of assuming that what would work in the fecund ground of his imagination must also work in the sterile ground of reality.

The Promise of Synthesis in "The Purloined Letter"

The final installment in the Dupin trilogy, "The Purloined Letter," concludes with a synthesis of Poe's analytical detective and Peirce's ideal rhetor, who creates forceful symbols by considering necessary conclusions (mathematics), reasoning through signs (logic), and creating beautiful images and forms (poetry). The case begins with a visit from the Prefect of police, who confides to Dupin that "Minister D—" is using an illicit letter to blackmail an unnamed lady for his own political advantage. The Prefect then relates in painful detail the thoroughness of his secret investigation of the Minister's apartments, grounds, and person as well as his frustration at not being able to find any trace of the letter despite having inspected every chair rung, bedknob, and paving stone with the most advanced microscopes available. Regarding the Minister's papers and books, the Prefect explains:

> We opened every package and parcel; we not only opened every book, but we turned over every leaf in each volume, not contenting ourselves with a mere shake, according to the fashion of some of our police officers. We also measured the thickness of every book-cover, with the most accurate admeasurement, and applied to each the most jealous scrutiny of the microscope. Had any of the bindings been recently meddled with, it would have been utterly impossible that the fact should have escaped observation. Some five or six volumes, just from the hands of the binder, we carefully probed, longitudinally, with the needles. (1978c, 981)

When the Prefect finally asks Dupin for his advice, Dupin tauntingly tells him to search again. Meanwhile, Dupin pays a visit to the Minister and immediately locates the letter—hiding in plain sight on the mantelpiece—and through a simple ruse steals it back, replacing it with a facsimile, and hands the original over to the Prefect, for a "small" fee of fifty thousand francs.

Once the elated Prefect has departed with the purloined letter, Dupin explains to his companion: "This functionary, however, has been thoroughly mystified; and the remote source of his defeat lies in the supposition that the Minister is a fool, because he has acquired renown as a poet. All fools are poets; this Prefect *feels*; and he is merely guilty of a *non distributio medii* in thence inferring that all poets are fools" (Poe 1978c, 986). Dupin's interlocutor asks if the Minister could really be a poet and adds, "The Minister I believe has written

learnedly on the Differential Calculus. He is a mathematician, and no poet"
(1978c, 986). Dupin corrects his companion for inferring a false dichotomy, in
this case making the logical error of mutual exclusivity in his reasoning. Dupin
explains that the Minister is actually both: "As poet *and* mathematician, he
would reason well; as mere mathematician, he could not have reasoned at all,
and thus would have been at the mercy of the Prefect" (1978c, 986). As Dupin
explains, "the mathematician argues, from his *finite truths*, through habit, as if
they were of an absolutely generally applicability—as the world indeed imagines them to be" (1978c, 987). That is, if the Minister had simply been a mathematician, he would have hidden the letter according to the hypothesis that a
secret thing of great value must be well hidden. As Dupin explains, "if the Minister had been no more than a mathematician, the Prefect would have been
under no necessity of giving me this check. I knew him, however, as both mathematician and poet, and my measures were adapted to his capacity, with reference to the circumstances by which he was surrounded" (1978c, 988). That is,
because the Minister was not only a mathematician but also a poet, he took into
account that his adversary, the Prefect of police, would search every possible
hiding place; therefore, "the Minister had resorted to the comprehensive and
sagacious expedient of not attempting to conceal it at all" (1978c, 990). This
shows that the Minister, like Poe's checkers player from "The Murders in the
Rue Morgue," considers his opponent and baits him into a trap.

Conclusion

Through Dupin, Poe valorized the power of analytical reasoning to develop a
rational narrative that neatly accounted for the facts in the case of the murders
in the Rue Morgue but failed to acknowledge the potential pitfalls of applying
these same procedures to the real-world case of Mary Rogers. Where abductive
reasoning and necessary conclusions play a role in argumentation, the poetic or
rhetorical must also play a role by taking into account the contingency of context, the character of audiences, and the practical judgments of lived experience
that undoubtedly influence persuasion. Because Poe failed to incorporate
poetic/rhetorical methods while inventing the argument of his early detective
narratives, he too was foiled when he attempted to solve the actual murder of
Mary Rogers. As a mathematical solution to the murder, Poe's fictional detective
provided an intellectually elegant hypothesis, but when further details about the

murders emerged, his narrative seemed as ridiculous as an Ourang-Outang with a shaving razor.

Though Poe blamed the newspapers for concealing vital evidence in the case of Mary Rogers, the real problem with solving a case through analytical reasoning alone is that, while useful in invention, analytical reasoning may lead us to absurd conclusions when it is applied to the messiness of real-world contexts. Poe articulates this point in "The Purloined Letter" when he has Dupin joke: "I never yet encountered the mere mathematician who could be trusted out of equal roots, or one who did not clandestinely hold it as a point of his faith that $x^2 + px$ was absolutely and unconditionally equal to q. Say to one of these gentlemen, by way of experiment, if you please, that you believe occasions may occur where $x^2 + px$ is *not* altogether equal to q, and, having made him understand what you mean, get out of his reach as speedily as convenient, for, beyond doubt, he will endeavor to knock you down" (1978c, 988). The joke is that the quadratic equation cannot be applied to a problem that does not deal with relations of form and quantity. While Poe is simply suggesting that altering the strictly regulated assumptions about reality within which mathematicians operate will drive them to physical violence, the implication within the context of "The Purloined Letter" is that the abductive reasoning that allows a mathematician to solve a quadratic equation must be combined with empirical verification if it is to yield rhetorical force for the detective.

Though the abductive reasoning and necessary conclusions that underlie mathematical reasoning may be a robust method for tackling seemingly unsolvable problems, Poe's forced revision of the story of Mary Rogers's murder illustrates how mathematical reasoning oversteps its bounds when it treats the world as a puzzle with neat parameters. If we can learn a lesson from Peirce and Poe about the usefulness of rhetoric in mathematics, therefore, it may be twofold: first, that training in abductive reasoning and necessary conclusions informed by rhetorical sensibilities is absolutely essential to be able to reconfigure facts in creative ways, to invent new narratives about the relationship between facts, and to identify causes beyond the popular imagination. Second, that even as we generate hypotheses and speculate on consequences, we must nonetheless keep in mind the essential fallibility of our abductions and retain a healthy rhetorical appreciation for the contingency of situations, the complexity of our audience, and the humility that is necessary for acting in a world in which we only ever have partial understanding.

References

Aristotle. 2007. *On Rhetoric: A Theory of Civic Discourse*. 2nd ed. Translated by George A. Kennedy. Oxford: Oxford University Press.

Brent, Joseph. 1994. *Charles Sanders Peirce: A Life*. Bloomington: Indiana University Press.

Buchler, Justus. 1939. *Charles Peirce's Empiricism*. New York: Routledge.

Bybee, Michael D. 1991. "Abduction and Rhetorical Theory." *Philosophy and Rhetoric* 24 (4): 281–300.

Colapietro, Vincent Michael. 2007. "C. S. Peirce's Rhetorical Turn." *Transactions of the Charles S. Peirce Society* 43 (1): 16–52.

Dipert, Randall R. 1977. "Peirce's Theory of the Geometrical Structure of Physical Space." *Isis* 68 (3): 404–13.

Domino, Brian. 1994. "Two Models of Abductive Inquiry." *Philosophy and Rhetoric* 27 (1): 63–65.

Hookway, Christopher. 2007. "Peirce on Icons and Cognition." In *Conceptual Structures: Knowledge Architectures for Smart Applications*, edited by Uta Priss, Simon Polovina, and Richard Hill, 59–68. Berlin: Springer.

Kragh, Helge. 2012. "Is Space Flat? Nineteenth-Century Astronomy and Non-Euclidean Geometry." *Journal of Astronomical History and Heritage* 15 (3): 149–58.

Lyne, John R. 1980. "Rhetoric and Semiotic in C. S. Peirce." *Quarterly Journal of Speech* 66:155–68.

Moriarty, Sandra E. 1996. "Abduction: A Theory of Visual Interpretation." *Communication Theory* 6 (2): 167–87.

Peirce, Charles. 1984. *Writings of Charles S. Peirce: A Chronological Edition*. Vol. 2, *1867–1871*. Edited by Peirce Edition Project. Bloomington: Indiana University Press.

———. 1998. *The Essential Peirce: Selected Philosophical Writings*. Vol. 2, *1893–1913*. Edited by Peirce Edition Project. Bloomington: Indiana University Press.

———. 2010. *Philosophy of Mathematics: Selected Writings*. Edited by Matthew E. Moore. Bloomington: Indiana University Press.

Perelman, Chaïm. 1982. *The Realm of Rhetoric*. Notre Dame: University of Notre Dame Press.

Perelman, Chaïm, and Lucie Olbrechts-Tyteca. 1971. *The New Rhetoric*. Notre Dame: University of Notre Dame Press.

Poe, Edgar Allan. 1978a. "The Murders in the Rue Morgue." In *The Collected Works of Edgar Allan Poe*, vol. 2, *Tales and Sketches*, edited by Thomas Ollive Mabbott, Eleanor D. Kewer, and Maureen C. Mabbott, 521–74. Cambridge, MA: Belknap Press.

———. 1978b. "The Mystery of Marie Rogêt." In *The Collected Works of Edgar Allan Poe*, vol. 3, *Tales and Sketches*, 715–88.

———. 1978c. "The Purloined Letter." In *The Collected Works of Edgar Allan Poe*, vol. 3, *Tales and Sketches*, 972–97.

———. 2008. *The Collected Letters of Edgar Allan Poe*. 3rd ed. Edited by John Ward Ostrom, Burton R. Pollin, and Jeffrey A Savoye. New York: Gordian Press.

Part 3

Mathematical Argument and Rhetorical Invention

6

Rhetoric and Mathematics in the Saturnian Account of Atomic Spectra

Joseph Little

In this chapter, I enlist Schiappa's threefold taxonomy (rhetoric *in* mathematics, rhetoric *of* mathematics, and mathematical language *as* rhetorical) as a heuristic to guide my analysis of the Saturnian account of atomic spectra advanced in 1904 by Hantaro Nagaoka, one of Japan's leading physicists. I argue that Saturn provided the underlying conceptual scheme for Nagaoka's discursive problem space, and in so doing licensed the axioms and subsequent lines of mathematical reasoning that would be seen by his audience as logically permissible in undertaking a solution. Interestingly, the Saturnian analogy functions substantively in Nagaoka's mathematics without the aid of natural language; the presence of analogy is discernible in the equations themselves. I contend that some of Nagaoka's equations function qualitatively in his arguments rather than quantitatively. At times, the equations indexically reference larger concepts and therefore can be thought of as a special kind of citation rather than as a tool for computation. Finally, by considering the ways in which Nagaoka's argument is situated within larger social and historical contexts, I uncover the fundamentally synecdochal character of Nagaoka's mathematical account of the atom, an understated part-to-whole relationship that places him at odds with British mathematical physicist G. A. Schott regarding the proper practice of science but not, to Schott's dismay, with the rest of the European physics community. As a historical case study, the chapter devotes considerable time to describing the intellectual landscape within which Nagaoka was working before discussing the relationships between rhetoric and mathematics latent in his work.

Nagaoka's Ruling Exigence: The Puzzle of Spectral Emissions

The scientific study of atomic spectra began with the discovery that a gas electrified in a cathode ray tube as well as certain solids placed in a flame emit not the

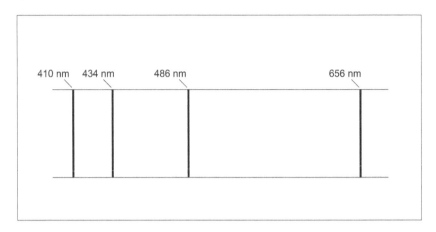

6.1 | Line spectrum for hydrogen in the visible region.

continuous Newtonian spectrum but a series of discrete, colorful lines when viewed through a diffraction grating. Under certain conditions, the lines appear black against a colorful spectrum. Under others, they appear bright and colorful against a black background. However, given a particular substance, the position and, therefore, the wavelength of the lines never change.

By the 1860s, scientists realized that each element in nature emits its own line spectrum, what J. R. Rydberg (1900, 207) would later call an element's "own characteristic and invariable language." Hydrogen and only hydrogen, for example, was known to have a visible line spectrum consisting of one red line at wavelength 656.210 nanometers, one greenish-blue line at 486.074 nanometers, and two violet lines at 434.010 and 410.120 nanometers respectively (fig. 6.1). With the knowledge that each element emits a unique spectral fingerprint, scientists could separate the light coming from a sample of interest to them—a star, a comet, an electrified gas, and so on—and accurately identify its chemical composition. Within a decade, scientists were publishing massive reports of spectral measurements to six significant figures (Conn and Turner 1965, 66). Precise as they were, however, the measurements were merely observational, leading many scientists to wonder whether they would ever discern an underlying order to the growing empirical record. Was there a pattern to the placement of the lines? And if so, which branches of mathematics were capable of expressing it?

The first breakthrough came in 1885 when J. J. Balmer realized that the wavelengths of the visible lines in the hydrogen spectrum could be described with astonishing accuracy by a simple formula:

$$\lambda = h \; \frac{m^2}{m^2 - 2^2}$$

In this equation, λ represents the wavelength of the emitted light in nanometers, h is an empirically derived constant (364.56 nanometers), and m is a stepwise variable that takes on positive integer values (Balmer 1885). When m is set equal to 3, 4, 5, and then 6, Balmer's formula yields wavelengths of 656.208, 486.080, 434.000, and 410.130 nanometers respectively, which agree with Anders Ångström's experimental measurements of the hydrogen spectrum (White 1934, 5). In that one formula, Balmer established the law-like behavior of spectral lines, at least in the case of hydrogen. Over time, scientists realized they could more easily see patterns and relationships in their data when line spectra were described in terms of frequency rather than wavelength. Balmer's formula was therefore soon rendered in the equivalent frequency form:

$$n = R \left(\frac{1}{2^2} - \frac{1}{m^2} \right)$$

In this equation, n represents the frequency of the emitted light in waves per centimeter, R is an empirically derived constant (109721 in $1/\lambda$ units), and m remains a stepwise variable that takes on positive integer values.

Given Balmer's success in predicting hydrogen's line spectrum, could a more generalized formula be found to predict the line spectra of all elements in the universe, thereby equipping scientists with a mathematical technique for identifying the chemical composition of any material under investigation? Five years later, J. R. Rydberg (1890) produced just such a formula, generalizing from Balmer's work to a simple formula describing the frequencies of all lines of all atomic spectra:

$$n = \frac{R}{(m_1 + \mu_1)^2} - \frac{R}{(m_2 + \mu_2)^2}$$

In this equation, n represents the frequency of the emitted light in waves per centimeter, R is an empirically derived constant (109721 in $1/\lambda$ units), m remains a stepwise variable that takes on positive integer values, and μ_1 and μ_2 are constants specific to the series and element under study. The development of this equation was a significant achievement for theoretical spectroscopy. Whereas Balmer's formula accounted for the line spectrum of hydrogen, Rydberg's predicted the line spectrum of every element in the universe, thereby providing the nascent field with one of its first covering laws.

Complicating matters, however, was the fact that atomic spectra were known to display not only lines but also thicker prominences called bands. In 1886, Henri Deslandres discovered a formula for the frequencies of the lines comprising the individual bands present in the spectra of nitrogen, cyanogen, oxygen, and water vapor. Like Balmer's, it was surprisingly simple:

$$n = a + bm^2$$

where n represents the frequency of the line, a is the frequency of the band head (the dense edge of the band), b is a constant associated with certain characteristics of the band under study, and m remains a stepwise variable that takes on positive integer values (fig. 6.2). And though it was less accurate than Balmer's, Deslandres calling it a "first approximation," it revealed much finer distinctions than Balmer's had and encompassed the regularities of more elements, four compared to Balmer's single case of hydrogen. Later, Deslandres showed that the same formula described not only the position of the lines within a single band (fig. 6.2, bottom), but also the position of all of the bands throughout an element's spectrum (fig. 6.2, top). In other words, the same quadratic relationship that governed the spacing of the bands within a band spectrum governed the spacing of lines within an individual band. If ever nature had shown itself to be orderly, it was here.

Although precision and rigor were the hallmarks of spectral analysis, what mattered most to theorists were the general qualities of the atom that could be deduced from the newly discovered laws. From Balmer's and Rydberg's formulas, theorists could see that the lines crowded more closely together as they converged on a shared limit, $R/4$ in Balmer's case. What is more, their frequencies were conspicuously restricted by the fact that they were always proportional to an expression involving ratios of small integers: $1/4 - 1/m^2$, where $m = 3, 4, 5$, and 6 in the case of hydrogen. Spectral lines never appeared at intervals between whole numbers like $m = 2.4$ or $m = 8.11$. Scientists had seen these patterns before, most clearly in musical harmonics. A taut string of a guitar or piano, for example, only generates tones at frequencies that are integer multiples of its

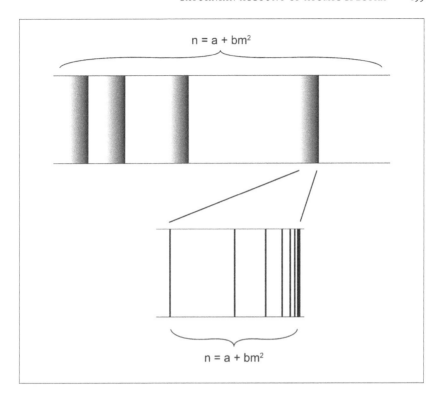

6.2 | Quadratic relationship among bands and among lines within a single band.

fundamental frequency or "first harmonic." The second, third, and fourth harmonics are twice, three times, and four times the fundamental frequency; their corresponding wavelengths are 1/2, 1/3, and 1/4 the wavelength of the fundamental frequency (table 6.1). When graphed, harmonic wavelengths produce the same crowding effect that scientists noticed in spectral lines. (Compare fig. 6.3 with fig. 6.1 to see the similarity.) They thus thought of spectral lines as having an essentially harmonic quality about them.

Table 6.1   Harmonics of taut string

Harmonic	Frequency	Wavelength
First	n	λ
Second	$2n$	$\lambda/2$
Third	$3n$	$\lambda/3$
Fourth	$4n$	$\lambda/4$
Fifth	$5n$	$\lambda/5$

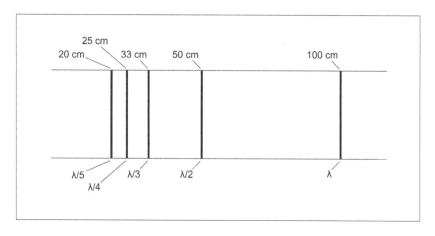

6.3 | Example of the crowding effect in first five harmonics of a taut string.

In Deslandres's formula for a single spectral band, theorists could see another quality of the atom. As with any quadratic equation—a polynomial equation of the second order—Deslandres's $n = a + bm^2$ revealed an arithmetic progression at work deep within the spectral lines of a band. When graphed by frequency, not only did the interval between successive lines decrease as the lines converged on the band head, but it also did so by a fixed amount (table 6.2, third column). It was not that the difference between the lines was constant, as was mistakenly implied in some textbooks, for that would have meant the lines were equidistant. Instead, it was the *difference in the difference* between the lines—to be clear, the $\Delta\Delta n$—that remained constant for a given series. The interval between the lines (Δn) then, and not the frequencies of the lines themselves (n), formed the arithmetic progression (table 6.2).

Table 6.2   Example of Deslandres's arithmetic progression ($a = 1\,000, b = 3$)

Frequency of line	Interval between lines (Δn)	Difference between intervals ($\Delta\Delta n$)
$n_m = 1 = 1003$		
$n_m = 2 = 1012$	9	
$n_m = 3 = 1027$	15	6
$n_m = 4 = 1048$	21	6
$n_m = 5 = 1075$	27	6
$n_m = 6 = 1108$	33	6

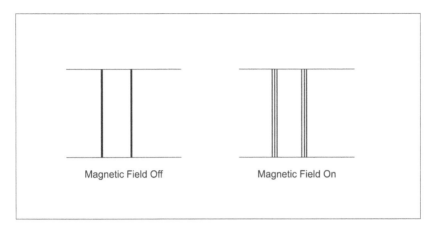

Magnetic Field Off   Magnetic Field On

6.4 | The Zeeman effect.

Around the same time, improvements in light-diffraction technology gave scientists an unprecedented look at some of the subtler features of the atom (Harrison 1958; Pais 1986). With the help of this more sensitive imaging, Pieter Zeeman (1897a; 1897b) discovered the remarkable phenomenon that now bears his name, the Zeeman effect: the broadening and splitting of spectral lines by a magnetic field (fig. 6.4). The implications were significant, for the fact that spectral lines but not bands were magnetically influenced suggested to theorists that line and band spectra were of different physical origins within the atom.

By the turn of the twentieth century, any theorist who could deduce from his theory of atomic structure a set of laws of motion that would manifest known spectroscopic qualities (e.g., harmonic distancing of spectral lines) while affording a plausible account of the Zeeman effect was thought to be well on his way to a true account of the atom, one that carried with it the empirical backing of three decades of Europe's highly prized spectroscopic research. It was to this ambitious end that Hantaro Nagaoka set himself in 1903.

Nagaoka's Response: A Saturnian Account of Spectral Emissions

Born in 1865, Nagaoka grew up in the tumultuous times following the Meiji Restoration, when feudal Japan was supplanted by a centralized government dedicated to developing a modern, techno-scientific Japan on par with the West. Pursuing this dream of a modern Japan, Nagaoka took his doctorate in physics in 1893 at the University of Tokyo, the premier institution of Western learning in the country, studying under the Scottish geophysicist C. G. Knott (Koizumi 1975). He then spent the mid-1890s in Europe studying under Herman Helmholtz and Max Planck in Berlin before moving to Munich to take courses with Ludwig Boltzmann. It was during his Munich days with Boltzmann that Nagaoka was introduced to the work of Scottish mathematical physicist James Clerk Maxwell. Maxwell's work made such an impression on the ambitious young scholar that he not only purchased a two-volume set of Maxwell's scientific papers, but also decided to follow Boltzmann, a committed Maxwellian, to Vienna when he was appointed there (Yagi 1964). Meiji Japan took all of Europe as its role model but especially Great Britain, home of the Industrial Revolution and victor over China in the First Opium War of the 1840s (Koizumi 1975). It comes as little surprise then that in his work on atomic structure Nagaoka (1904b) drew heavily from Maxwell's 1859 Adams Prize–winning essay, "On the Stability of the Motion of Saturn's Rings."

Maxwell had offered a masterful analysis of the dynamics of Saturn and its rings, concluding reductio ad absurdum that the rings must consist of a large number of separate particles all orbiting the planet at high velocity. Further, Maxwell (1983, 93) had concluded that satellites orbiting in a circular, flat ring can endure minor displacements in three directions without jeopardizing the overall stability of the system. In Maxwell's work on Saturn's rings, Nagaoka found what he thought might be a potentially fruitful model for thinking about the movement of electrons around the atom. Reasoning by analogy, Nagaoka mapped the three directions of displacement onto the electrons of his ideal atom:

DISPLACES RADIALLY (external force, electron)
DISPLACES ANGULARLY (external force, electron)
DISPLACES NORMALLY (external force, electron)

Once imported into the atomic domain, these three analogical correspondences defined the ways an electron could vibrate, or be displaced from its orbit, when influenced by an external force. Although vibrations in the radial direction failed to evince any spectral implications, Nagaoka's deduction of the frequency of electronic vibration in the normal direction, that is, perpendicular to the plane of the orbit, produced striking results. Just as Maxwell (1983, 96–97) had arrived at

$$n = \pm\sqrt{S+\frac{R}{\mu}J} \text{ with } J = \Sigma\left(\frac{\sin^2 m\theta}{2\sin^3\theta}\right)$$

where n represents the frequency of normal vibration, S and R/μ are constants, and the series term J takes on increasing discrete values as $m = 1, 2, 3, \ldots$, Nagaoka (1904b, 448–50) arrived at the analogous formulation for his ideal atom:

$$n = \pm\sqrt{S+\mu J} \text{ with } J = \Sigma\left(\frac{\sin^2 h\theta}{3\sin^3\theta}\right)$$

Here again n represents the frequency of normal vibration, S and μ are constants—though of electrostatic rather than gravitational force—and the series term J takes on stepwise values as $h = 1, 2, 3$, and so on. In textbook hypothetico-deductive style, Nagaoka then sought out a candidate atomic configuration that would connect his theoretical frequency equation to the body of spectroscopic data amassed by experimental physicists across Europe. Nagaoka realized that by simply assuming that the atom's ring was composed of a relatively large number of orbiting electrons, he could expand J in a way that allowed him to rewrite his frequency equation as $n = a + bh^2 + ch^3 + \ldots$, where $h = 1, 2, 3, \ldots$ and a, b, and c are constants. This revision resulted in Deslandres's quadratic formula for band spectra ($n = a + bm^2$), including its arithmetic progression and its harmonic crowding of lines once the small values of ch^3 and so on were neglected. It was a point of profound agreement between theory and observation, one Nagaoka was quick to recognize: "The interval between successive frequencies decreases as h is increased. Constructing the frequency lines as functions of h, we find a close resemblance with the band-spectrum. . . . In fact, the above equation is but an extension of Deslandres' empirical formula in a slightly altered form" (Nagaoka 1904b, 449–50).

Turning next to angular motion, Nagaoka deduced another frequency equation that admitted spectral implications (1904b, 451):

$$\pm n = \frac{\omega}{\sqrt{a+bh^2+ch^4+\ldots}}$$

where n is the frequency of angular displacement, $h = 1, 2, 3, \ldots$, ω is the angular velocity of the particles orbiting the central particle, and a, b, and c are constants. When graphed, the equation produced the same kind of crowding effect that scientists had observed in spectral lines since the days of Balmer (fig. 6.1). Nagaoka expressed in prose what anyone who laid eyes on his graph already knew: "The frequency increases as h is increased, and the nature of the series shows that the spectral lines corresponding to these vibrations will gradually crowd together when h is large. The qualitative coincidence of the above result with the line-spectrum is at once evident, if h be not small" (Nagaoka 1904b, 451). Nagaoka interpreted this patterned crowding of lines as more than coincidental. There, in the pages of the *Philosophical Magazine*, he had deduced from Saturnian axioms the essential mathematical qualities of band and line spectra that physicists had known experimentally for decades, and on that basis he concluded it was indeed vibrations of electrons in the normal direction—in and out of the plane of revolution—that caused the appearance of bands in an element's spectrum, whereas vibrations in the angular direction—the periodic speeding up and slowing down of electrons in their orbits—were responsible for the lines.

That the normal and angular electronic vibrations existed in different planes and therefore did not interfere with each other, provided Nagaoka with the final result he needed: a plausible account for the mysterious fact that line but not band spectra exhibited the Zeeman effect. In other words, scientists knew that the dynamics of line and band spectra must be independent of each other since only the former exhibited the Zeeman effect. The Saturnian geometry imported into Nagaoka's theory provided a qualitative, mechanical account of just such a dynamic independence.

Rhetoric *in* Nagaoka's Mathematics: How Analogy Can
Inform Mathematics

The movement of material in Saturn's rings provided the base for Nagaoka's analogy, licensing the axioms and subsequent lines of mathematical reasoning that would be accepted by his audience as a viable explanation of the Zeeman effect. As such, Nagaoka's 1904 paper serves as a clear example of rhetoric operating in mathematics. As Schiappa rightly contends in chapter 2 of this collection, "no branch of mathematics can get off the ground without acceptance of a large number of stipulated definitions; this is conspicuously true of axiomatic-deductive approaches to proofs, which are the approaches most often thought of as free of rhetoric." In Nagaoka's case, the analogical correspondences structuring the problem space provided the very "stipulated definitions" he needed to bootstrap (in the Quinean sense) his mathematics. Nagaoka then proceeded by way of abductive reasoning in much the way Johannes Kepler and Nagaoka's hero, Maxwell, had reasoned in their famous works (Carey 2009; Gentner 2002; Nersessian 1992).

Given the base domain of the Saturnian analogy, Nagaoka's work is deeply shaped by what Richard Engnell calls "primary materiality" (1998, 4): the non-symbolic motivations of human discourse that are located at the most rudimentary levels of biological and physical processes. Drawing on Kenneth Burke's (1966) notion of "sheer animality" and its relationship to human symbolicity, Engnell (1998) carefully distinguishes this primary materiality from the secondary materiality of technological, economic, and sociopolitical structures. It is this latter form, he argues, that is more commonly taken up by rhetoricians. In Nagaoka's case, however, it was by way of the Saturnian planetary system, a primary materiality, that he was justified in proposing a nuclear atomic theory—a theory of the atom that included a massive central particle akin to Saturn. Similarly, it was by way of the Saturnian analogy that Nagaoka found nonformal justification for proposing rings of electrons in high-velocity orbits rather than the slow-moving or even static particle systems of his contemporaries, who subscribed to different mathematics given their different material and symbolic commitments. Therefore, while the puzzle of the atom's curious spectral emissions provided the exigence for Nagaoka's theory, it was Saturn and its action upon its rings that guided the mathematics he would employ, providing the context for proper reasoning. In other words, the Saturnian analogy—that is, the set of structural relationships that Nagaoka mapped from Saturn to his ideal atom—created much of the *orthos logos*, or "right reason," engendering Nagaoka's mathematical discourse (Little 2001; 2008). As such, it stands as a palpable example of one of the ways material relationships can provide conceptual anchoring for mathematical reasoning to take place.

That is not to say, however, that Nagaoka's symbolic activity was subservient to a materialist anchoring. In abductive reasoning, analogy provides a starting point, but the mathematical statements that arise from subsequent deductions ultimately adjudicate the veracity of that starting point. The Saturnian analogy afforded Nagaoka nonformal justification for positing his initial equations of electronic motion, analogically derived from Maxwell's precedent. However, it was his ability to deduce from those equations the atom's known spectral behavior that certified the Saturnian hypothesis in the end.

Finally, it should not go without notice that Nagaoka's work shows us an example of analogy operating substantively in mathematically driven scientific argument without the aid of natural language. In *Motives for Metaphor in Scientific and Technical Communication*, Timothy Giles (2008, 83) examines the role of the solar system analogy in modern theorizing of the atom. Giles argues that the analogy formally emerged in 1902 in the work of Oliver Lodge and remained viable until it was abandoned by Niels Bohr in 1913. Giles's exploration of the solar system analogy is deft and richly contextual; however, it refrains from examining the mathematics in question, choosing instead to focus on natural language expressions such as "orbit" and "path" as well as extended natural language comparisons between the solar system and the atom. In doing so, he assumes that analogy, serving constitutively in scientific texts, would necessarily present itself in the natural language portion of the argument. In Nagaoka's case, however, we can see Maxwell's equation for the frequency of the normal oscillation of the satellites about Saturn,

$$n = \pm\sqrt{S + \frac{R}{\mu}J} \text{ with } J = \Sigma\left(\frac{\sin^2 m\theta}{2\sin^3\theta}\right)$$

functioning analogically in Nagaoka's 1904 work in which he introduces this equation

$$n = \pm\sqrt{S + \mu J} \text{ with } J = \Sigma\left(\frac{\sin^2 h\theta}{3\sin^3\theta}\right)$$

to represent the vibration of his theoretical electrons orbiting in the normal plane. In the very syntax of the mathematics, Nagaoka's audience would likely have recognized Maxwell's work, despite the fact that he is not mentioned in the pages surrounding the equation. Nagaoka's case is not unusual. It is often in the mathematics, as distinct from the natural language prose of an argument, that scientists see and evaluate semblances, insights, intellectual debts, and logical inferences that shape the adjudication of scientific claims.

Rhetoric *of* Nagaoka's Mathematics: How Mathematics
Can Inform Science

Schiappa's category of rhetoric *of* mathematics inspires the question "How
might Nagaoka have deployed mathematics to enhance the credibility and
persuasiveness of his scientific argument?" One way of understanding the persuasive force of mathematics in his work would be to ask another, more opera-
tionalized question: "How did Nagaoka know when to stop deducing more
equations from his initial set of Saturnian assumptions?"

To explore this question, I draw on Charles Sanders Peirce's icon-index-
symbol terminology, a semiotic framework that Alan Gross (2006; 2008) has
used for visual exegesis. "An icon," explains Gross, "is a sign that depicts; a
photograph or a drawing of a microbe is an icon" insofar as it physically
resembles its referent (2006, 449). An index, on the other hand, relates to its
referent by way of causation or indication. In the case of causation, an index
can be thought of as a form of evidence since the sign is present only when its
referent is. Gross offers a Geiger counter reading as an index of radiation
because the latter causes a predictable response in the former. Last in Peirce's
triad is the symbol, a sign related to its referent by neither physical resemblance nor causation or indication, but rather by way of social convention.
The name "Chris," the equals sign "=," the number "4," and the Apple computer
logo are all examples of symbols.

In Nagaoka's case, we can say that some of his mathematical expressions
functioned indexically in his scientific argument. In the case of line spectra, he
stopped deducing further mathematical statements when he reached an equation whose graph qualitatively resembled the crowding effect of spectral lines
observed in laboratories across Europe and America and catalogued in Hein-
rich Kayser's *Handbuch der Spectroscopie* (Heilbron 1964, 112). Just as Geiger
counter hits signify the presence of ambient radiation, the crowding-line graph
deduced from Nagaoka's mathematics signified a fundamental quality of his
Saturnian model—the harmonic distancing of spectral lines—which was also
known to exist in the empirical atom. Although the graph was not conclusive—
since Nagaoka could not rule out the possibility that a false theory might generate this same quality of spectral crowding—it was nonetheless an instance of
profound theory-data fit. In this way, Nagaoka engaged in the abductive logic of
a diagnostician rather than in a process of inductive or deductive reasoning

(Geertz 1973), his crowding-line graph serving indexically as one of his primary epistemic waypoints along the path to a presentable scientific argument.

In the case of band spectra, once again Nagaoka was guided by a mathematical cue, this time not by an equation whose graph accorded with nature but by an equation itself. As he proceeded from his Maxwellian equations of electronic motion, he stopped only when he had roughly deduced Deslandres's formula for band spectra:

$$n = a + bm^2$$

In the context of Nagaoka's work, this equation never took on a computational role. It was never enlisted in a calculation, estimate, or quantitative comparison because Nagaoka was working deductively from Saturnian assumptions toward a qualitative account of the atom. Instead of taking on a computational role, the arrival at this mathematical equation communicated a single meaning: that the quality of arithmetic progression that was known to exist within the atom was also evident within Nagaoka's atomic model. As such, Deslandres's formula took on an indexical function in Nagaoka's inquiry and subsequent oral and written communications identical to that of the graph of crowding lines. It served as evidence of profound theory-data fit as Nagaoka methodically worked from mathematical assumptions to a scientific argument whose claims resounded with the experimental facts available at the time.

The field of citation studies has argued that in-text citations are used not only to give credit to sources and to enable readers to locate texts, but also to serve a variety of rhetorical ends, including aligning the writer with a particular discourse community and establishing the writer's authority by way of familiarity with the relevant research. In some cases, citations stand in for concepts themselves, argues Small (1978), the way "(Bazerman 1988)" can signify a particular approach to the study of scientific discourse as distinct from, for example, the approach symbolized by the citation "(Gross 1992)." In Nagaoka's case, Deslandres's equation likewise stands in for a concept—that of arithmetic progression—and therefore, hermeneutically, it functions in ways similar to that of a citation. But it also serves indexically as evidence of the presence of that quality in the set of mathematical relationships from which it was deduced, for only systems that are harmonic-arithmetic in nature are likely to yield Deslandres's equation. In short, Nagaoka's argument shows that a mathematical equation can function indexically, symbolically, and qualitatively in a given case without taking on a computational role.

More specifically, it is the quadratic form of Deslandres's equation that served as the source of meaning for Nagaoka and his audience, for quadratics (the family of equations that can be expressed in the form $ax^2 + bx + c = 0$) entail the arithmetic progression and harmonic crowding of spectral lines that the atom was known to emit. What is important here is not the individual quantities that can be substituted for variables but the syntax of the expression, which is why Nagaoka chose to present the cleaner, though less computationally accurate, form by omitting the trailing terms ($+ cb^3 + \ldots$). Although Nagaoka's treatise is mathematically rigorous, his is a qualitative argument in the sense that it is the category of relationships between variables (rather than any numerical value resulting from calculation) that matters, and that is why it is the various visual signs of the relationships between the fundamental qualities of his ideal atom, which resemble those of the empirical atom, that hold the persuasive force of the equations for him and his readers.

Nagaoka's Mathematical Language *as* Rhetorical: Mathematical Conventions in Context

To more fully understand Nagaoka's perspective on the scope of his Saturnian theory, one particular feature of the larger context of inquiry must be discussed. This feature is revealed in a debate between Nagaoka and his harshest critic, the British mathematical physicist G. A. Schott. Nagaoka never thought that his Saturnian theory would provide a full account of the atom, but rather thought that it served as a useful mathematical model of one of the atom's most prominent features, spectral emissions.

Upon reading Nagaoka's work, Schott (1904a) fired off a letter to *Nature* claiming that he had investigated a Saturnian analogy five years earlier and had abandoned it because it produced an unstable atom. In the plausible scenarios he investigated, the electrons would vibrate out of control, leading to the annihilation of the atom itself. In short, Schott argued forcefully that Nagaoka's Saturnian atom was not nearly stable enough to be held up as a candidate theory for the true atom. In response, Nagaoka (1904a) suggested that Schott had misunderstood him: the Saturnian account did not consider all of the electrons in an actual atom. Rather, it was simply a mathematical idealization that enabled physicists to draw upon Maxwell's mathematics to consider the dynamic properties of only those electrons in the outermost orbit. It is for this reason that I argue that the entirety of Nagaoka's classic paper of 1904 must be seen as an instance of synecdoche, an argument in which "a part of something is used to signify the whole" (Abrams 1999, 98). Nagaoka's goal was, after all, to provide a mathematical account of known spectral regularities much more than it was to offer a full rendering of the atom. Whatever the behavior of the rest of the electrons might be, whether orbital or static or something else altogether, it fell far beyond the scope of Nagaoka's purpose. Therefore, much of Schott's criticism, Nagaoka argued, missed the point. Nagaoka then closed his letter by directing the reader's attention back to the promising spectral findings of his partial and idealized atom, including his mathematical deduction of Deslandres's formula for band spectra.

A somewhat exasperated Schott continued the volley, claiming that Nagaoka had seriously understated the instability of the hypothesized atom. After correcting a few technical errors in Nagaoka's treatment, he again calculated a series of scenarios that would lead to disturbances entirely incompatible with the notion of a stable atom, even given Nagaoka's hedged criterion of "quasi-stability." Schott did concede Nagaoka's point that if one conveniently neglected the vast majority of intra-atomic electrons and decided to make the central particle arbitrarily charged, the Saturnian atom (now understood as synecdochal) "can certainly be made stable"—"but is this allowable?" he asked incredulously (1904b, 387). Here we see a palpable example of Schiappa's contention that mathematics, like any symbol system, operates within sociohistoric contexts of established practice where the communal acceptance of novel symbolic deployment is anything but assured. In Schott's estimation, Nagaoka's mathematics failed to underwrite his scientific argument because the geometry selectively addressed parts of the atom (those relevant to the study of spectral emissions) at the expense of the whole, a move Schott found entirely beyond the conventions of acceptable scientific practice. It was his final comment on the matter.

Unfortunately for Schott, the rest of the physics community was more accepting of Nagaoka's synecdoche. In 1905, despite the Schott-Nagaoka debate, E. C. C. Baly acknowledged Nagaoka in his London-based publication *Spectroscopy*: "It has been suggested that each atom of a chemical element consists of a central positively charged mass with the system of negatively charged electrons in motion round about it, very similar to the ring system of the planet Saturn with electrical substituted for gravitational attraction. Nagaoka has shown that it is possible to account for both band and line spectra by vibrational disturbances occurring in such a system" (qtd. in Yagi 1972, 75). In *The Value of Science*, the eminent Henri Poincaré likewise singled out Nagaoka for his promising attempt to explain the spectral qualities of the atom. Discounting Schott's critique, he writes:

> *Electrons and Spectra.* These dynamics of electrons can be approached from many sides, but among the ways leading thither is one which has been somewhat neglected, and yet this is one of those which promise us the most surprises. It is movements of electrons which produce the lines of the emission spectra. . . . Why are the lines of the spectrum distributed in accordance with a regular law? These laws have been studied by the experimenters in their least details; they are very precise and comparatively simple. . . .
>
> . . . A Japanese physicist, M. Nagaoka, has recently proposed an explanation; according to him, atoms are composed of a large positive electron surrounded by a ring formed of a very great number of very small negative electrons. Such is the planet Saturn with its rings. This is a very interesting attempt, but not yet wholly satisfactory; this attempt should be renewed. (Poincaré 2001, 311–12)

And, surely to Schott's dismay, Ernst Rutherford (1911) himself, in one of the most famous papers in physics, acknowledged Nagaoka as a progenitor of modern atomic theory for his work on the Saturnian atom. Whatever else we may take from Nagaoka's story, we can be sure that the success of his mathematical account relied in part on the strategic use of synecdoche in a discourse community that welcomed such referential partiality as an accepted practice within science.

Conclusion

By closely reading Nagaoka's treatise with Schiappa's threefold taxonomy in mind, we find that mathematics is not operating exclusively beyond the purview of rhetoric but with its help. In his work, Nagaoka used the Saturnian analogy to provide the concepts, geometries, and lines of reasoning he needed to justify the application of his mathematics as he endeavored to solve the mystery of the atomic origins of spectral emissions. Drawing from the language of the rhetorical tradition, we can say that Nagaoka's materialist analogy created much of the *orthos logos* supporting the choices of mathematical discourse by suppressing the candidacy of some mathematical choices while justifying others (Little 2001; 2008).

Nagaoka's treatise, while pervasively and rigorously mathematical, was ultimately qualitative in nature. In the case of line spectra, his derivation of an equation whose graph qualitatively resembled the crowding effect of spectral lines certified his Saturnian theory as fundamentally harmonic, a quality also known to exist in the atom. In this way, his mathematics led to an instance of profound theory-data fit that all but his harshest critics valued. Nagaoka likewise derived from his Saturnian analogy an approximation of Deslandres's empirically derived formula for band spectra. The fact that Nagaoka omitted trailing terms from this mathematical equation attests to my claim that his arrival at Deslandres's formula was not valued for its computational utility. Rather, it was valued because it indexically signified that his "ideal" atom shared an important quality with the empirical atom: the arithmetic progression of lines comprising a spectral band. Accordingly, we can see Deslandres's formula operating in Nagaoka's argument as a special kind of in-text citation since it stood in for the larger concept of arithmetic progression, an association his audience would have immediately recognized.

Nagaoka's treatise also serves as an example of analogy functioning substantively in scientific argument without being present in the natural language of the text. In the very syntax of his mathematical description of the dynamics of electrons about the nucleus, Nagaoka's audience would have seen the Saturnian analogy figuring prominently in his claims despite the fact that Maxwell is not mentioned in the accompanying natural language. This example encourages scholars interested in the contribution of analogy to mathematically driven scientific argument to include mathematical expressions in future rhetorical analyses. No longer can we confine ourselves to the prose and visuals of a technical argument without opening ourselves to legitimate concerns over the adequacy of our analytic gaze.

Finally, Nagaoka's entire argument can be interpreted as an example of synecdoche, which sponsored the most direct and vociferous challenge to his work. As G. A. Schott argued, Nagaoka, in accounting for spectral emissions, had also advanced an atomic theory that he knew was partial and incompatible with any full rendering of the true atom. This outraged Schott, who felt that Nagaoka's "ideal atom" was no atom at all but rather a convenient theoretical contrivance that could never exist in nature. Strikingly, Nagaoka conceded this point, and yet he advanced his Saturnian theory nonetheless on the grounds that his ideal atom, contrived or not, accounted for spectral emissions. In the Nagaoka-Schott debate, then, we have a dispute that does not contest the nature of the atom as much as it reveals the profound incommensurability between Nagaoka and Schott on the question of what constitutes good science.

References

Abrams, M. H. 1999. *A Glossary of Literary Terms.* Boston: Heinle and Heinle.

Balmer, J. J. 1885. "Notiz über die Spectrallinien des Wasserstoffs." *Naturforschende Gesell-schaft, Basel, Verhandlungen* 7:548–60.

Burke, K. 1966. *Language as Symbolic Action.* Berkeley: University of California Press.

Carey, S. 2009. *The Origin of Concepts.* New York: Oxford University Press.

Conn, G. K. T., and H. D. Turner. 1965. *The Evolution of the Nuclear Atom.* London: Iliffe Books.

Deslandres, H. 1886. "Spectre du pole negatif de l'azote: Loi generale de repartition des raies dans les spectres des bandes." *Comptes Rendus* 103:375–79.

Engnell, R. A. 1998. "Materiality, Symbolicity, and the Rhetoric of Order: 'Dialectical Biolo-gism' as Motive in Burke." *Western Journal of Communication* 62:1–25.

Geertz, C. 1973. *The Interpretation of Cultures.* New York: Basic Books.

Gentner, D. 2002. "Analogy in Scientific Discovery: The Case of Johannes Kepler." In *Model-Based Reasoning: Science, Technology, Values,* edited by L. Magnani and N. J. Nerses-sian, 21–39. New York: Kluwer Academic / Plenum.

Giles, T. D. 2008. *Motives for Metaphor in Scientific and Technical Communication.* Amityville, NY: Baywood.

Gross, A. G. 2006. "The Verbal and the Visual in Science: A Heideggerian Perspective." *Sci-ence in Context* 19:443–74.

———. 2008. "The Brains in *Brain:* The Coevolution of Localization and Its Images." *Journal of the History of the Neurosciences* 17:380–92.

Harrison, G. R. 1958. "The Controlled Ruling of Diffraction Gratings." *Proceedings of the American Philosophical Society* 102:483–91.

Heilbron, J. L. 1964. "A History of the Problem of Atomic Structure from the Discovery of the Electron to the Beginning of Quantum Mechanics." PhD diss., University of Cali-fornia, Berkeley.

Koizumi, K. 1975. "The Emergence of Japan's First Physicists, 1868–1900." *Historical Studies in the Physical Sciences* 6:3–108.

Little, J. 2001. "Toward Sociocultural Sensitivity in Rhetorical Studies of Analogy: Theoreti-cal and Methodological Considerations." *Journal of Technical Writing and Communica-tion* 31:257–66.

———. 2008. "The Role of Analogy in George Gamow's Derivation of Drop Energy." *Techni-cal Communication Quarterly* 17:220–38.

Maxwell, J. C. 1983. "On the Stability of the Motion of Saturn's Rings." In *Maxwell on Saturn's Rings,* edited by S. G. Brush, C. W. F. Everitt, and E. Garber, 68–158. Cambridge, MA: MIT Press.

Nagaoka, H. 1904a. "A Dynamical System Illustrating the Spectrum Lines." *Nature* 70:124–25.

————. 1904b. "Kinetics of a System of Particles Illustrating the Line and the Band Spectrum and the Phenomena of Radioactivity." *Philosophical Magazine* 7:445–55.

Nersessian, N. 1992. "How Do Scientists Think? Capturing the Dynamics of Conceptual Change in Science." In *Cognitive Models of Science*, edited by R. Giere, 3–44. Minneapolis: University of Minnesota Press.

Pais, A. 1986. *Inward Bound: Of Matter and Forces in the Physical World*. New York: Oxford University Press.

Poincaré, H. 2001 [1905]. *The Value of Science*. New York: Modern Library.

Rutherford, E. 1911. "The Scattering of Alpha and Beta Particles by Matter and the Structure of the Atom." *Philosophical Magazine* 21:669–88.

Rydberg, J. R. 1890. "Recherches sur la sonstitution des Spectres d'émission des éléments chimiques." *Kungliga Vetenskaps Akademiens Handlinger* 23:1–155.

————. 1900. "La distribution des raies spectrales." *Congrès International de Physique, Rapports* 2:200–224.

Schott, G. A. 1904a. "A Dynamical System Illustrating the Spectrum Lines and the Phenomena of Radioactivity." *Nature* 69:437.

————. 1904b. "On the Kinetics of a System of Particles Illustrating the Line and Band Spectrum." *Philosophical Magazine* 6:384–87.

Small, H. G. 1978. "Cited Documents as Concept Symbols." *Social Studies of Science* 8:327–40.

White, H. E. 1934. *Introduction to Atomic Spectra*. New York: McGraw-Hill.

Yagi, E. 1964. "On Nagaoka's Saturnian Atomic Model (1903)." *Japanese Studies in the History of Science* 3:29–47.

————. 1972. "The Development of Nagaoka's Saturnian Atomic Model, II (1904–05)." *Japanese Studies in the History of Science* 11:73–89.

Zeeman, P. 1897a. "Doublets and Triplets in the Spectrum Produced by External Magnetic Forces." *Philosophical Magazine* 44:55.

————. 1897b. "On the Influence of Magnetism on the Nature of the Light Emitted by a Substance." *Philosophical Magazine* 43:226–39.

7

The New Mathematical Arts of Argument | Naturalistic Images and Geometric Diagrams

Jeanne Fahnestock

The wings of the human mind are arithmetic and geometry.
—Philip Melanchthon, 1536

In 1695 the *Philosophical Transactions* published an account of the dissection of a scallop by Martin Lister, a physician who contributed many "observations" on spiders, shells, fossils, and much else to the Royal Society's inaugural journal (Unwin 1995, 209). Lister's Latin text was accompanied by a detailed engraving (fig. 7.1), the work of one of his talented teenage daughters who illustrated their father's extensive publications on conchology (Roos 2012, 32). Typical for the time, both Lister's description and the illustration deploy the conventions of geometry, verbal and visual, to construct the familiar scallop as an object of scientific knowledge. In both the main text and the italicized figure legend that follows it, geometrical terms are used to identify the parts encountered in the dissection: interior surfaces are described as concave (*concava*) or flat (*planae*), parts are oriented to one another at right angles (*ex rectis angulis*) or obliquely (*oblique positorum*), cavities are like circles (*veluti circuli*), and protuberances are orbicular (*orbicularis*). And one depicted part is described as shaped like a rhomboid (*figura rhomboide*).

Well known to everyone with Lister's level of education, the rhombus, rhomboid, and trapezium are distinguished in book 1, definition 22 of Euclid's *Elements* as four-sided figures lacking four right angles but equilateral in the first case, having parallel pairs of equal sides in the second, and having neither constraint in the third. And indeed the area marked *dd* in the illustration and also described as rhomboidal in the legend (*rhomboidos*) is a distinct four-sided shape with parallel top and bottom, closer to what today would be called an isosceles trapezoid. Perhaps the depiction of *dd*, done under Lister's supervision, was eased into resembling the familiar geometrical figure, a figure typically found in the opening pages of geometry texts in the sixteenth and seventeenth centuries (fig. 7.2) and the subject of important proofs in book I of the *Elements* (1482, 1.34, 1.35). Contemporary drawings and photographs of scallop anatomy show no such structure.

There is also another layer to the intersection of the natural and geometrical in Lister's text, for the Latin *rhombus*, taken from Greek geometry (ῥόμβος), was also the name of a diamond-shaped flatfish, the rhombus (e.g., turbot).

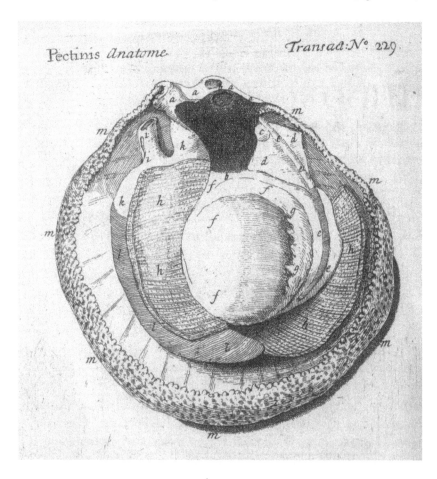

7.1 | Anne Lister's engraving of a scallop from *Philosophical Transactions*, no. 229 (1695), frontispiece. Republished with permission of The Royal Society, London; permission conveyed through Copyright Clearance Center, Inc.

32 *Rhombus(or a diamonde)is a figure hauing foure equall fydes, but it is not rightangled.*

Definition of a
Diamond figure

This figure agreeth with a fquare, as touching the equallitie of lines, but differeth from it in that it hath not right angles, as hath the fquare. As of this figure, the foure lines *A B, B C, C D, D A*, be e-quall, but the angles therof are not right angles. For the two angles *A B C* and *A D C*, are obtufe angles, greater then right angles, & the other two angles *B A D* and *B C D*, are two acute angles leffe then two right angles. And thefe foure angles are yet equall to foure right angles: for, as much as the acute angle wanteth of a right angle, fo much the obtufe angle excedeth a right angle.

C.i.   *Rhombaides*

The firſt Booke

Definition of a
diamondlike fi-
gure.

33 *Rhombaides (or a diamond like)is a figure, whofe oppofite fides are e-quall, and whofe oppofite angles are alfo equall, but it hath neither e-quall fides, nor right angles.*

As in the figure *A B C D*, all the foure fides are not equall, but the two fides *A B* and *C D*, being oppofite the one to the other, alfo the other two fides *A C* and *B D*, being alfo oppofite, are equall the one to the o-ther. Likewife the angles are not right angles, but the angles *C A B*, and *C D B*, are obtufe angles, and op-pofite and equall the one to the other. Likewyfe the angles *A B D*, and *A C D*, are acute angles, and oppo-fite, and alfo equall the one to the other.

Trapezia or
tables.

34 *All other figures of foure fides befides thefe, are called trapezia, or tables.*

Such are all figures, in which is obferued no equallitie of fides nor angles: as the figures *A* and *B*, in the margŕt, which haue nei-ther equall fides, nor equal angles, but are defcribed at all aduen-ture without obferuation of order, and therefore they are called irregular figures.

Definition of
Parallel lines.

35 *Parallel or equidiftant right lines are fuch, which being in one and the felfe fame fuperficies, and pro-duced infinitely on both fydes, do neuer in any part concurre.*

As are the lines *A B*, and *C D*, in the example.

7.2 | Rhombuses from Henry Billingsley, *The Elements of Geometry* (London: Imprinted by John Daye, 1570), fols. Cir–Civ. Courtesy of the Library of Congress.

And indeed, a few sentences before invoking the geometrical figure, Lister describes two circular cavities in the image as resembling the eyes of the rhombus fish (*rhombi piscis*).

On the image itself, the conventions of geometric diagramming are noticeable in the multiple letters used to mark distinct parts of the scallop's anatomy. Such letters derive from the labeling practices in geometry texts, and the specific technique of using multiple lower-case letters to mark contiguous areas was common in human anatomy texts in the sixteenth and seventeenth centuries, prototypically in Vesalius's *De humani corporis fabrica* (1543, 608). Such letters allow a more complex integration between the text and the image than would be possible without them. Parts distinguished by letter labeling are also deliberately engraved as distinct through means such as dissimilar crosshatching, and this visual analysis, though it seems simply given, is in fact an arguable construction, something that Lister himself was aware of since he referred to such contributions as "arguments" (Unwin 1995, 216). Indeed, in the accompanying text, Lister often expresses uncertainty about what he is observing, and in the conclusion to his brief account of "the parts of this animal" (*partium huius animalis*) he notes that any attempt to identify their uses is futile unless they can be explained on the basis of analogy (*ex analogia*) with other animals of the same genus. The term *analogia* here means something more mathematical than a mere comparison, as does the goal of identifying the integral parts of the whole animal. Both analysis and analogy had a formal, well-defined place in the intellectual tool kit of the seventeenth-century natural philosopher, the result of a new art of argument in the making for more than a century.

An overview of the mathematically inflected dialectical/rhetorical art of argument that natural philosophers like Lister would have been familiar with is the subject of this chapter. Changes in the sixteenth century are key. Humanist educational reforms brought a renewal to all the arts of the trivium and quadrivium and an interanimation among them that is difficult to appreciate from a twenty-first-century perspective that separates the mathematical and verbal. But mathematical standards and methods of reasoning were incorporated into the sixteenth-century verbal art of dialectic to a surprising extent, and as a consequence, geometrical ways of seeing diagrammatically were combined with a new attention to and facility in naturalistic representation and image reproduction. These tactics of arguing with words and images, established in the sixteenth century, were taken up by later generations of scholars who studied the "book of nature" as a second scripture, using divinely ordained tools to reveal divinely ordained meanings.

Tracing the sources of these changes first requires sampling the tradition that led most directly to texts like Lister's: the educational reforms instituted by the Reformation in northern Europe, where the changes happened precipitously. The era's newly inclusive art of argument is on display in one of the most widely used textbooks on dialectic of the sixteenth century, Philip Melanchthon's *Erotemata dialectices* (Dialectical questions).[1] A detailed examination of the place of mathematics in this text reveals the attention to quantity as a property, to mathematical demonstration as a standard of reasoning also applicable to moral and practical argument, and to lines of argument derived from geometry and arithmetic. At the same time, and somewhat surprisingly, the text promotes visualization as a method of definition, reflecting the new accuracy in printing images then available. Visualization brought with it the conventions of geometrical diagramming and arguing, a source of new methods of coming to terms with the biological world. These changes can be sampled in the works of natural history in the sixteenth and seventeenth centuries, culminating in more fully mathematical approaches to the natural world in the nineteenth century.

The New Synthesis

Beginning in the 1520s, the reformers of the University of Wittenberg changed the educational landscape of the Protestant territories and cities under their influence. With the advice and sometimes the direct intervention of Luther's co-reformer Philip Melanchthon, university curricula were altered, secondary schools reformed or created, and the teaching staff at both levels turned over to younger teachers from approved universities, often from Wittenberg itself. Coming into prominence in these changes was mathematics in all four of its quadrivium forms: arithmetic, geometry, astronomy, and music. From the early 1520s, Wittenberg's revised statutes called for two professors of mathematics at the university, and in 1536 the two installed were Georg Joachim Rheticus to teach the "lower mathematics" of arithmetic and geometry, and Erasmus Reinhold to teach the "higher mathematics" of astronomy, astrology, and geography (K. Reich 1993, 110–13). Both men were key facilitators of Copernican astronomy, Rheticus by stimulating the production of *De revolutionibus* and writing a preliminary account, the *Narratio prima*, and Reinhold by producing the *Prutenic Tables*, calculating astronomical positions according to the Copernican model (Westman 1975).

As part of these reforms at both the Latin school and university level, Protestant printers issued textbooks whose appropriateness was often signaled by a preface from Melanchthon himself. Among these introductory mathematics texts were Georg Peurbach's *Elementa arithmetices* (1534; written in the fifteenth century), Johannes Vögelin's *Elementale geometricum* (1536), and a new edition of Sacrobosco's venerable *De sphaera* (1531), the standard introductory astronomy text.[2] Perhaps the most mathematically important text to have the patronage of a preface from Melanchthon was Michael Stifel's *Arithmetica integra* (the 1544 edition), seen as a landmark text in algebra for introducing notation for multiple unknowns in problems (Heeffer 2010, 65–69).

Melanchthon himself was trained in mathematics under Johannes Stöffler during his years at the University of Tübingen (1512–1518), but he did not add to the substance of the mathematical treatises he prefaced. Instead, his introductions stressed the crucial importance of mathematics for every scholar and for every Christian. In the preface to Vögelin's *Elements of Geometry*, for example, he comments that "the philosopher needs geometry, for the beginnings of natural philosophy take their origin from there" and "no one without a knowledge of that art understands what is the power of demonstrations" (Melanchthon 1999, 98). Melanchthon also wrote or collaborated on academic orations that were encomia to mathematics. Perhaps his highest praise comes in an oration of 1536, delivered by Rheticus but cowritten by Melanchthon, which contains the famous observation that "the wings of the human mind are arithmetic and geometry"[3] (Melanchthon 1963a, 288).

In the intellectual landscape of the sixteenth century, unlike the intellectual landscape of today, areas of learning were connected territories rather than separated islands, and the grounding discipline of all the arts was dialectic, a general art of argumentation and therefore of reasoning, applicable to all disciplines—medicine, law, theology, astronomy, natural philosophy, and even mathematics. Giovanna Cifoletti has made the case that the combination of dialectic with rhetoric and mathematics in the sixteenth century produced a general "art of thinking" (2006, 371, 379). Indeed, Cifoletti goes further and makes the stronger claim that "the reform of mathematics and science in the sixteenth century was first conceived of as a reform of the 'art of thinking'" (2006, 371). Cifoletti's emphasis, however, is on the work of Ramus in both his *Dialectique* of 1555 and his *Scholaraum mathematicarum* of 1569. Offered here is a different view of the dialectic/mathematics overlap based on Melanchthon's widely disseminated textbook, the *Erotemata dialectices* (first edition 1547, last expanded edition, 1555). An examination of Melanchthon's dialectic, the product of forty years of work on successive versions and revisions (Mack 1993, 319–33), will help to uncover the productive overlap between the discursive and the mathematical arts in the sixteenth century. Written in the form of questions with extended answers, this advanced text was printed in at least forty-six editions into the seventeenth century (Mack 1993, 320), and Melanchthon's status in the Reformation solidified its widespread use in Protestant schools and universities. On one side, it clearly reflects Melanchthon's view that dialectic was inseparable from rhetoric with its focus on everyday reasoning in deliberative, forensic, and epideictic situations. Melanchthon in fact also wrote three treatises on rhetoric, and in the last of these, the *Elementorum rhetorices* of 1532, he claims that "such is the connection of Dialectic and Rhetoric, that their difference can barely be discovered" (Melanchthon 1532, A5iii). Invention and disposition are the same in both arts, he explains, as rhetors are accustomed to using the places of invention taught in dialectic.[4] The examples that fill the pages of the *Erotemata dialectics* on civil, interpretive, and practical matters could just as easily have come from one of his rhetoric texts.

On the other side, less well known, Melanchthon's dialectic is also continuous with the arts of arithmetic and geometry using demonstrations or proofs. Connections between his rhetorical/dialectical art of arguing and mathematics are on display in the very opening pages when he sets up the definition of dialectic and answers the question "About what things is Dialectic concerned?" as follows: "It concerns all the subject matters or questions that people should be taught about, just as Arithmetic is concerned with all the things that should be numbered. Indeed, there is an important affinity between Dialectic and Arithmetic. God imparted a knowledge of numbers into natural intelligence so that the mind distinguishes things. . . . Moreover, after Arithmetic has numbered things, Dialectic enters and attributes different names and definitions to distinct things; it investigates numbers, parts, causes, effects, and other related things. Then in reasoning it combines the coherent and distinguishes the diverse"[5] (Melanchthon 1548). Melanchthon reinforces the connection by explaining that just as arithmetic "at one time numbers gold, at another stones, at another other things" (*Arithmetica numerat alias aurum, alias lapides, alias alia* [1547, biiv]), so dialectic is concerned with "any matter whatsoever" (*circa quamlibet materiam*

[biiv]), making both arts completely general in their application. Geometry, too, is given a founding role in dialectic when Melanchthon observes that "Aristotle constructed Dialectic in great part from Geometry" (*Extruxit autem Aristoteles dialecticen magna ex parte geometria* [1548, 67r]).[6] Altogether, Melanchthon believed that all these foundational arts, dialectic teaching definition and distinctions, arithmetic teaching computation, and geometry teaching measurement (1548, 36v), were, along with other innate beliefs, divinely ordained: "God however handed down Arithmetic, Geometry and certain things in Natural Philosophy and Ethics so that they would govern many parts of life and would be supports for Divine teaching."[7]

Mathematics in the *Erotemata dialectices*

The detailed role of mathematics throughout Melanchthon's dialectic is best explicated by following the overall organization of his textbook, an organization typical of his time. The text is divided into four books that sensibly and fully partition a complete art of argument, beginning with the categories of words in book I as they are used to form statements and definitions. Book II then identifies types of statements and book III their combination into argumentative structures. Finally, the sources for inventing arguments (topics) and judging their cogency (fallacies) are presented with numerous examples in book IV.

Book I of Melanchthon's *Erotemata dialectices* includes the parts of Aristotle's *Organon* concerned with the basics of predication. Under the five *predicables* (genus, species, differentia, property, accident) and the ten *predicaments* (the term used to refer to Aristotle's ten categories),[8] the text covers the semantic basics of statement formation, and the book concludes with sections on Definition and Division, the foundational intellectual tasks of dialectic. A defining statement, for example, can predicate the genus of a species or one of its properties or accidents. So, for example, in "Happiness is a pleasurable motion of the heart," a claim that is more physiological than moral for Melanchthon, *motion of the heart* is the genus and *pleasurable* a property (1548, 28v; "Laeticia est suavis motus cordis").

Among the predicaments (categories for properties or accidents), the second in importance after substance was Quantity, and in Melanchthon's definition of this category of predicates, the overlap between dialectic and the mathematical arts is especially clear: "Quantity is the magnitude of a thing or the number, and it is appropriate for a quantity to be measured or divided" ("Quantitas est magnitudo rei vel numerus. Et quantitati proprie convenit mensurari vel dividi" [1548, 18v]). Number, he continues, accompanies anything whatever, and magnitude is inseparable from corporeal substance. He then refers students to arithmetic and geometry as the arts that deal fully with Quantity and says he will only briefly review the nomenclature of a few species. As a "first division" of Quantity, he repeats the distinction between continuous and discrete quantities. Continuous quantities express the magnitude of a body, namely its length, width, and depth ("Alia est quantitas continua, alia discreta. Continua est magnitudo corporum, videlicet, longitudo, latitudo & profunditas" [1548, 19r]). Continuous quantity in turn has three "species," line, plane, and solid [*Linea, Superficies, and Corpus*], and Melanchthon's understanding of the purely cognitive status of these geometrical entities is clear: "Line is length without width and depth, whose extremes are two points. Moreover young people learn that this definition is not asserted here about depicted lines but about length alone, abstracted from bodies, that should be contemplated in the mind, just as the Mathematician considers quantities without matter"[9] (Melanchthon 1548, 19r). In contrast to continuous quantities, a discrete quantity is defined as number ("Discreta est numerus"),[10] and number in turn is defined as a "multitude assembled from units, such as *two, three, four*" (1548, 19r). Melanchthon will repeat this distinction in a section on the qualities of different disciplines when he stresses again that arithmetic is for computation and geometry for measurement (1548, 36v).[11]

Melanchthon also gives three interesting "properties" of the overall category Quantity in arguments. First, numbers have no efficacy in themselves and cannot cause anything; hence they cannot be used in superstitious rites. Second, they lead to the relations *equal and unequal* (for numbers) and *same and not the same* (for geometrical figures that are or are not congruent), and these relations can be extended to comparisons among other entities. But third, quantities do not accept *more* and *less* (*magis et minus*) but *greater* and *smaller* (*maius et minus*). While this distinction seems odd (and does not survive in English), Melanchthon wants to reserve the adverbial *more or less* formula for qualities that come in degrees and the adjectival *greater and smaller* as verbal labels for relations among quantities. Much of natural philosophy in the next century will be concerned with finding quantitative ways of expressing qualities like *hot* and so undoing this distinction.

Under the predicament of Quality, Melanchthon identifies (after habits, natural powers, and emotions) a fourth type, namely *figures*, meaning the geometrical figures of circle, triangle, rectangle, etc. (1548, 28v). Thus to describe the shape of something (as Lister frequently does) is to state one of its qualities. In this division of the category Quality, Melanchthon follows Aristotle (Barnes 1984, I:16), but he also claims to be using the language of mathematicians, who, he says, call qualities *poiotetes*. He acknowledges that recent writers dispute this assignment, but he maintains that it is still worthwhile "to speak with those most clever and learned men, the ancient Geometers" (1548, 28v; "loqui cum ingeniosissimis & doctissimis hominibus veteribus Geometris"). Again, all this discussion is included in a general treatise on argumentation.

After book I covers the functional roles of different terms within categorical statements, book II covers the four basic types of these statements—*All x is y, No x is y, Some x is y, Some x is not y*. In later sections of this second book, Melanchthon also covers the relationships and manipulations among these types, such as conversion, the "flipping" of a proposition by exchanging extremes to create in some cases an immediate inference from an equivalent proposition: *No sugar is bitter / Therefore no bitter thing is sugar* (1548, 66v). In defending the usefulness of teaching these verbal rearrangements, he cites their importance both in geometry and in the way numbers are talked about. He follows with what we would now consider an example of the commutative law in arithmetic: "*Four times five are twenty / Therefore five times four are twenty*. Here you see an example of conversion. In the same way in Geometry, many things are concluded by conversion."[12] And indeed Euclid 1.27 and 1.29 prove converse propositions about alternate angles created when parallel lines are intersected by a third line.

Picking up from the statement types discussed in book II, book III next covers the arrangements of statements into inference structures, beginning with the syllogism in its three figures and twenty-four modes. In justifying why three figures are necessary, Melanchthon explains that those learned in geometry will recognize the correspondence between the reasoning structures represented in the three figures of the syllogism (reasoning genus to species, species to genus, and separating species) with their own methods: synthesis is congruent with the first figure, analysis with the third, and arguments from the impossible with the second. This characterization of the methods of geometrical reasoning comes not directly from the *Elements* but from Pappus of Alexandria's commentary, widely known in the sixteenth century,[13] and Melanchthon repeats this distinction in book IV: "For Geometers, the usual and most well-known terms [are] composition or synthesis which proceeds *a priori*. The reverse [is] resolution or analysis, which returns *a posteriori* to principles."

Melanchthon's understanding of valid inference forms encompasses more than the categorical syllogism, however. He includes the enthymeme, induction, example, and sorites—all familiar forms in rhetoric but also listed here in his expanded dialectic—as well as a form that has since disappeared: the expository syllogism. This form follows the third figure, in which the middle term is the subject of both the major and the minor premise. But in this variant, the middle term stands for some individual entity that is literally under the eyes or even in the hands of the arguer: *This thing [haec res] heats the ventricle / This thing [haec res] is zinzibar [ginger] / Therefore zinzibar heats the ventricle* (Melanchthon 1548, 89r).[14] Thus Melanchthon's discussion of the inferences licensed by this particular argument form includes visual encounters with an object, and oddly, he extends the application of the expository syllogism to include not just encounters with one thing but with commensurate things, citing in justification the first of the Common Notions in the *Elements*:

> Some inferences derive their rationale from this geometrical principle: *Whatever things are equal to a third thing are equal among themselves.* This *sententia* [principle], since it is known by nature, confirms the inference about which it speaks. It is moreover like the following inference:
>
> > *This line is equal to this three-fourths of a foot,*
> > *And this second line is equal to the same three-fourths.*
> > *Therefore, these are all equal among themselves.*
>
> Geometrical examples are clear: And since in relation to these it is clear that the inference is valid, it will indeed be valid in other matters. Thus on account of similarity the geometrical principle is applied here (1548, 89v–90r).[15]

Melanchthon continues by warning, however, that this geometrical principle does not apply to dissimilar things or to inequalities or unequal terms. One cannot say, for example, that *a man is a biped* and *a hen is a biped* and therefore that *a man is a hen* since these two entities are so dissimilar to begin with.

Book III concludes with still more inference structures, namely the hypothetical, copulative, and disjunctive syllogisms, and here again Melanchthon uses geometrical examples to clarify distinctions. He is especially concerned about a certain phrasing typical in Greek that involves two consecutive negative questions. He finds this form of phrasing characteristic of geometry as in the

propositions that he identifies as I.8 and III.24 in Euclid, and he quotes the latter as follows:"*Are not chords equal and are not arcs equal*, that is: It is not consistent when chords are equal for arcs not to be equal" (1548, 116r).[16]

Book IV of the *Erotemata dialectices*, which is as long as the first three books
combined, turns to the sources of the actual arguments filling the forms defined
in book III. But first Melanchthon surveys the territory of argument by identifying three types of syllogisms, though this time not in terms of their forms as
in book III but in terms of their content and argumentative force, namely the
demonstrative, dialectical, and sophistic. The demonstrative syllogism "consists
of subject matter that is necessary and unchangeable" (1548, 119r ["constat ex
materia necessaria & immota"]), the dialectical is based on "probable matters"
(1548, 119r ["constat ex materia probabili"]), and the sophistic uses falsehoods
that have an "appearance of truth" (1548, 119v ["speciem veri"]). He goes on to
explain that in the dialectical genus, "some [arguments] are more probable and
evident, some less. Since, however, they are not unchangeable demonstrations,
as are those in geometry, they are called dialectical arguments. For so they were
called in antiquity. It assigned demonstrations to Geometry; other arguments
on matters in natural philosophy and morals for the most part it called dialectical. For although some among these [arguments in natural philosophy and
morality] are demonstrations, nevertheless much the greater part contains only
probable matters"[17] (1548, 119v). But now Melanchthon has a nomenclature
problem, for the same term, *dialectical*, is used both for probable arguments *and*
for the entire art that includes both demonstrative and probable reasoning. So
he clarifies in concluding this section: "Now a part of dialectic is also the doctrine about demonstrations" (1555, 228 ["Nunc dialecticae pars est etiam doctrina de demonstrationibus"]). Thus *dialectic* is a hyponym labeling probable
arguments, and it is also a hypernym, a term for the overall completely general
art of argument covering all forms and degrees of certainty and applicable to all
subjects.

There is no question that for Melanchthon the certainty achieved in demonstration is not limited to mathematics but can be achieved in arguments on
any subject, so long as they too start from fixed definitions and axioms. Melanchthon distinguishes three sources of these certainties: universal experience
(e.g., that fire is hot), inborn principles, and correctly ordered reasoning.
To begin with the third source, certainty can come from correctly formed
syllogisms:

And because among the foremost syllogisms are the syllogisms of arithmetic, we should examine in these the clarity of inferences. Proportions are known from nature. In a combination of these in the true order of a syllogism, given three numbers, the fourth follows necessarily. *One pound of wax is worth one drachma, / Here there are ten pounds, / Therefore the price will be ten drachmae.* This is the order of the first figure of the syllogism, if you adapt it skillfully. For the major premise is set for a universal: *A pound of something costs one drachma.* Then, however many pounds there are, there are so many drachmae. Moreover, just as proportion, which can be judged by the natural light of the mind, shows that the inference in such arithmetical syllogisms is valid and immutable, so in certain syllogisms the inference is observed to be valid and immutable from demonstrations that are not summoned from distant things but from the obvious and evident.[18] (1548, 123v)

These "obvious and evident" sources included inborn principles, further divided into the speculative and practical. Speculative principles are "propositions known from nature from which physical and mathematical teachings are extracted, such as *The whole is greater than any of its parts; Things equal to another thing are equal among themselves; Of one simple body, the natural motion is singular [unicus]* (1548, 124r); the first two here are Euclidean common notions,[19] and Melanchthon identifies these clear and basic principles "in geometry" as *koinas ennoias kai axiomata* (common concepts and axioms) (1548, 124v). Unlike these speculative principles, practical principles are propositions known from nature that should control human behavior, and "just as arithmetical and geometrical principles and the demonstrations made from them are firm and immutable, so we know in fact that practical principles and the demonstrations made from them are firm and immutable. Just as the assent is firm and perpetual with which I assent to this proposition, *2 times 4 are 8*, so it is right for humans to embrace this proposition with firm assent: *Do not commit theft.* But in this darkness, we see that divine light less, and depraved emotions weaken assent in practical matters"[20] (1548, 124v–125r). In Melanchthon's discussion of the fallacy of amphiboly toward the end of book IV, he again yokes the certainty of *twice 4 is 8* with similar innate principles that "have their own place in the mind and stay the same: *An honorable and tranquil life is a good thing; death is a bad thing*"[21] (Melanchthon 1555, 381).

Melanchthon's blending of demonstrative and probable reasoning into a single art of argument continues in the section on the topics of invention that dominates book IV of the *Erotemata dialectices*, and since these topics offer direction for constructing and not just analyzing arguments, they are perhaps the most critical part of this text for assessing how Melanchthon incorporates mathematical reasoning.[22] Among his list of twenty-eight lines of argument, two of the standard topics of invention overlap with, if they do not also derive from, geometry. These are the arguments *ab absurdo* (1548, 157v) and *ab impossibili* (1548, 158v) that are featured in Euclidean proofs.[23] In the *Elements* itself, reductio arguments involve adopting as though it were true a proposition that is the opposite of the one actually being proved. The arguer goes through with a construction based on the contradictory proposition and discovers that it leads to an absurdity. For example, proposition 6 of book I claims that in a triangle with two equal angles, the opposite sides will be equal. This is proved by imagining that one of the sides is less than the other and then constructing an internal triangle from this smaller side that would nevertheless have its other sides and angles equal to the original triangle, meaning that the contained triangle would be equal to the containing triangle or the less equal to the greater, an absurdity. In his adaptation of this topic, Melanchthon offers as examples of absurdities the Stoic and Epicurean doctrines on the status of the emotions as merely opinions and on the equal preferability of life and death (1548, 157v). The formula *ab impossibili* also appears in many of the proofs concerning the circle in book 3 of the *Elements*, and these also involve attempted constructions leading to contradictory outcomes, usually involving the less equaling the greater. Melanchthon's examples include the impossibility of any power other than the divine raising the dead or of heavy objects rising without external force (1548, 159r).

For arguments in natural philosophy, perhaps the most important inventional topic related to image practices is not a new but an enduring one, *De toto et partibus*, "On the Whole and the Parts" (1548, 142v–144v). In Euclid's *Elements*, a part-whole axiom is expressed in the fifth Common Notion, *The whole is greater than the part*, and elsewhere, as, for example, in the definitions preceding book VII. In addition, the figures used throughout the *Elements* are necessarily understood in terms of their parts (e.g., angles and sides). When Melanchthon uses the whole-part relation as an inventional device in verbal arguments outside mathematics, he identifies arguments based on some division of a whole, whatever the subject. The resulting parts identified in an arguer's division, he explains with his own division, can be distinguished by whether they are more or less principal, that is, whether their absence destroys the whole or not. Not surprisingly, it is the discipline of anatomy that gives Melanchthon his examples here: in the human body, the hair and tongue are less principal, the head, heart, and chest, etc., are more principal since without them the whole ceases to exist. Principal parts are further divided into the *substantial* (related to substance; e.g., body and soul are the substantial parts of a human) and the *integral*. Integral parts can "have a name from quantity or number, as when we partition a mass or divide numbers; for example, when a six-foot log is divided into two-foot long parts, those parts are called integral" (1543, 143r).[24] This clear numerical example is followed by others involving "relative" or mutually constituting parts of conceptual wholes; thus the parts of a trial, of conversion, and of baptism are also described as integral. From this characterization of part-whole relations, Melanchthon derives two rules of reasoning: *Given a whole, it is necessary to posit all the principal parts*, and *Destroy a principal part and the whole is destroyed* (1548, 143v–144r).

The relation of parts to a whole is altogether general and applies to objects and concepts as well as to numbers and figures. But a more direct connection to mathematics occurs in a new topic that Melanchthon adds, *From Proportion:*[25] "It pleases me very much to mention proportion among the places of invention," he writes, "for after substance and causes have been discussed, it is often also necessary to speak about quantity, about numbers, and about the comparison of numbers, since in daily life mention of costs is common and necessary. For example, the paterfamilias should see to it—indeed all lovers of justice should see to it—that the magnitude of a price does not exceed the available resources. This comparison is derived from the doctrine of proportions" (1548, 164v).[26]

Melanchthon divides his topic of proportion into two parts. The first, named *logos*, is, he says, also called *ratio*. It concerns naming the juxtaposition of two quantities of the same type of entity, and he offers as an example that "*of four things to two things the proportion is double*" ("*quarternarii ad binarium dicitur esse proportio dupla*") (1548, 164v). Melanchthon calls the second type of argument involving proportion *analogia*. This type concerns the similitude of two ratios, and he illustrates it with a numerical example: *4 is to 2 as 6 is to 3*. It is this understanding of *analogia* as juxtaposed proportions that Lister has in mind in the passage cited above.

In his explanation of proportion as a locus of invention in verbal arguments, Melanchthon stresses that dialectic cannot really include the mathematical arts where proportion is taught in detail. But he points out that subjects of argument involving numerical proportion do occur frequently, and he offers as an extended example an attempt to translate the amount of water changed into wine at Cana into German volume equivalents:

> When in Cana of Galilee, Christ ordered six jars [*hydrias*] to be filled with water, which he converted into wine, if it is asked how much wine he will present to the bridegroom according to our measures, analogy will show the answer. One hydria would contain two *metretas* and a *metreta* truly would contain thirty measures that in our terms we call *kannen*; therefore two *metretas* contain a measure that we now commonly call an *eymer*.[27] Thus just as one hydria has two *metretas*, so six hydrias have 12 *metretas*. Therefore he will present an amount of wine that we would call six *eymer*, which was no meager gift.[28] (1548, 165r)

Melanchthon had multiple potential sources for his mathematical version of the topic of proportion. First, book V of Euclid's *Elements* introduces the concept of *ratio* in definition 1 (*A ratio is a sort of relation in respect of size between two magnitudes of the same kind*; Heath translation [1952]), and of *proportion* in definitions 5 and 6 (definition 6: *Let magnitudes which have the same ratio be called proportional*; Heath translation [1952]).[29] The fact that Melanchthon uses the terms *logos* and *analogia* for ratio and proportion suggests that he was consulting Simon Grynaeus's first Greek edition of Euclid, published in 1533, an edition based on new Greek manuscripts circulating in Europe (De Risi 2016, 597). Otherwise, ratio and proportion were well represented in arithmetic books at the time with terms derived from earlier Latin editions of the *Elements*, namely the terms *proportio* (for *ratio*) and *proportionalitas* (for *proportion* or *proportionality*; on this terminology see Murdoch [1963, 238n1]). Under these terms, the topic was treated in Pacioli's popular *Summa de arithmetica, geometria, proportioni et proportionalita* (1494; written in Italian) and in his *De divina proportioni*, a treatise on the golden ratio famous for its illustrations by Leonardo Da Vinci. Another potential source for Melanchthon was the Latin arithmetic text circulated from Wittenberg, Peurbach's *Elementa arithmetices*, to which Vögelin's treatment of proportion was added as a fourth book. But a more immediate source for him for the 1547 *Erotemata dialectices* may have been the advanced text of Michael Stifelius, the *Arithmetica integra* of 1544, mentioned above, which includes an extensive account of specially named ratios including 4 to 3, the *sesquitertia* and 9 to 5 the *superquarparticus quintas* (Stifelius 1544, 49r–49v).

Melanchthon no doubt had treatments like Stifelius's in mind when he referred to fuller coverage of ratio and proportion available in the dedicated study of mathematics.

Visualization and Dialectic

The invocation of mathematics, as described above, occurs throughout Melanchthon's *Erotemata dialectices*. Mathematics provides a standard of demonstrative argument for dialectic, and its structures of reasoning, such as those using absurdity or proportionality, can also be adopted for probable arguments on any subject matter. How does this inclusive art of argumentation then connect with visualization as the opening example from Lister suggests? There are several points of continuity between the early modern dialectic tradition and the increasingly dominant visualization practices in sixteenth- and seventeenth-century natural philosophy, from the use of diagrams in dialectic textbooks themselves to the designation of images as a form of definition.

First, the diagrammatic visualizations in dialectic itself should be noted. Just as geometrical reasoning is prominent in dialectic, so are geometric diagrams since dialectic textbooks themselves actually used such diagrams. The most famous of these is the well-known square of oppositions used to illustrate the relationship among the four types of propositions: the universal affirmative, universal negative, particular affirmative, and particular negative. This use of the square with its four equal sides argues visually for the equal standing of the four statement types, and the potential inferential relationships among them are indicated by the connecting diagonals and sides.[30]

In the Byzantine tradition, another geometrical system of diagramming arguments appeared in manuscript dialectics,[31] and a much later record of this tradition of diagramming is preserved in early printed dialectical textbooks that remain close to the scholastic tradition. For example, in Johannes Eck's *Elementarius dialectice* of 1517, the section describing the standard forms of the syllogism illustrates the three "figures" of the syllogism with three geometrical figures (fig. 7.3).

The figure with crescents or *lunules* represents the first and most famous of the twenty-four forms known under the mnemonic *Barbara*, whose three "a's" represent three universal affirmative propositions.[32] In Eck's example, the three propositions are *Every oblivious person is incapable of study / Every drunk is*

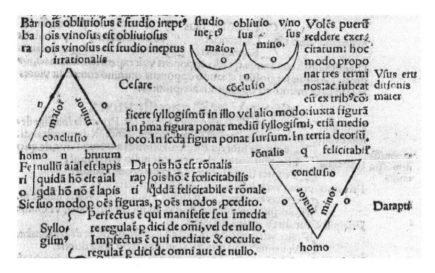

7.3 | Three "figures" of the syllogism with three geometrical figures, from Johannes Eck, *Elementarius dialectice* (Augustae Vindelicorum: Miller, 1517), Civ4. Courtesy of the Bayerische Staatsbibliothek.

oblivious / Therefore every drunk is incapable of study. The two half moons represent the motion from the major premise to the minor premise; the bottom half circle represents the overall inference, a conclusion from major premise by way of the minor. The equilateral triangle on its base represents *Ferio* in the second figure (e.g., *No animal is a stone; Some man is animal; Therefore some man is not a stone*), and the triangle poised on its apex represents *Darapti* using the third figure (*All men are rational; All men are capable of happiness; Some rational men are capable of happiness*). The explanation of the three figures in the text concerns the placement of the middle term in the diagram: it is in the middle for *Barbara*, at the top for *Ferio*, and at the bottom for *Darapti*. A teacher could clearly use the action of drawing these figures, as drawing was used in teaching geometry, to represent the structure of the inferences in each type of syllogism.

Geometric diagrams did then serve as illustrations of dialectical doctrine, but Melanchthon also generally linked knowledge with visualization of which geometrical diagrams were one source. This linkage is clearest in a passage added to the 1560 edition of the *Erotemata dialectices*, where *intellectual habits* are defined as knowledge and knowledge in turn is defined as "firm impressions of images in the brain" ("firmae impressiones imaginum in cerebro"). Archimedes is offered as an example of someone with greater knowledge because he has "firmer impressions of geometrical images in his brain than does Midas" (1560, 40).[33]

But what role, if any, did naturalistic rather than diagrammatic images have in dialectic and in image-based knowledge? Actually, an unusual feature in Melanchthon's *Erotemata dialectices* is the new importance he gave to naturalistic images as a means of definitional knowledge and argumentation. We noted above that book I of Melanchthon's dialectic is concerned with the semantic basics of predication and definition. But in the section "On Definition," following his extended discussion of the predicables and predicaments, Melanchthon distinguishes between defining words and defining things.

What is the definition of a thing?

The definition of a thing is a statement [*oratio*] that explains the essence or the causes or the parts or the accidents of a thing. And the difference between the definition of a word and the definition of a thing is easy to understand in this way: it is the definition of a word when you interpret a word from a foreign language with a more familiar word from our language and you name the genus, as when you say: *Centaury is a plant, which we call "tausent gulden," or Aurin.* You hear the genus and the name as less strange, and yet it can happen that the thing itself is still unknown. But if the plant is brought forward and placed under your eyes so that you can observe it, now you have a lucid definition. For the ancients said, and it is worth remembering: *All intuitive knowledge [intuitiva noticia] is definition.*[34] (Melanchthon 1548, 47v)

The neo-Latin adjective *intuitivus* modifying *knowledge* in this aphorism derives from *intueri*, meaning *to look at, to regard.* A defining knowledge of things can come then from seeing them, and it is not surprising that Melanchthon illustrates this point with the contemplation of a plant. When the *Erotemata dialectices* was published there were three ways of bringing an object such as a plant "under the eyes"—by having it present physically, by reading a verbal description of it (meeting the rhetorical standard of "bringing under the eyes"), or by looking at a naturalistic image of it such as those newly available in the printed illustrated herbals by Otto Brunfels (1530) or Leonhart Fuchs (1542).

One form of knowledge derived from natural images connects back to Melanchthon's new topic on proportion, and the source was one personally known to Melanchthon: Albrecht Dürer (Kuspit 1973, 177). Dürer wrote three works, originally in German, that were essentially on geometrical proportion applied to different depiction practices. The first, the *Unterweysung der Messung* (Teaching of measurement) of 1525, is a basic text on geometry and on principles of perspective for depicting geometrical figures and architectural objects. The second work covered architecture, especially for fortifications (1527), and the third was actually a compilation of Dürer's earlier writings on the proportions of the human body, published together shortly after his death as *Vier Bucher von menschlicher Proportion* (1528). All these works were translated into Latin by Melanchthon's friend Camerarius, and Melanchthon knew both the texts and their author. The most important treatise for the identifying parts of naturalistic objects and their mutual proportions is the last. There Dürer illustrated partitions of various body types, male and female, and sought the mathematical relations among the parts: length of head to length of body, forearm to whole arm, and so on (Schaar 1998). As a result of his studies, he departed from the notion of a single standard of proportional beauty for the human body (the goal of Da Vinci's Vitruvian man) and advocated instead the quantitative description of varying body types (Panofsky 1955, 262–67).

Two Traditions of Imaging: Diagrammatic and Naturalistic

Melanchthon's mid-sixteenth-century dialectic then inherits two traditions of imaging—the diagrammatic and the naturalistic. The diagrammatic tradition had of course the stronger connection with argumentation through its long association with geometry. Indeed, the oldest examples of "diagram plus text" in the Western tradition are fragments of Euclid,[35] and both the earliest surviving full manuscript of Euclid, dated to 888 CE (Bodleian MS D'Orville 301), and the first printed edition of Euclid in Latin in 1482 (fig. 7.4) combine text with images marked by letters, allowing the diagrams to be read in terms of the verbal directions or stages in the proof. This cooperation of text and diagram is essential since in traditional Euclidean geometry inferences are made directly from the diagram as well as from the verbal text (Manders 2008, 80, 88).

The conventions of geometrical diagrams were thus well established from late antiquity. What of the parallel tradition of representing things seen in the world? The ability of artists to convince viewers that they are seeing lifelike, even trompe l'oeil, images of small-scale objects was also achieved in classical Greece,

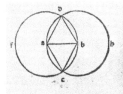

rus crefcit in infinitū:ficut magnitudo in infinitum minuitur.

Propofitio prima.

Riangulum equilaterum fupra datam lineam rectam col locare.

⟨ Efto data linea recta.a.b.volo fuper ipfam triangulū equilaterū cōftituere fuper altera eius extremitaté.f.in puncto a. ponam pedē circini immobilem:⁊ alterū pedem mobilem extendam vfq3 ad.b.⁊ defcribā ſm quantitatem ipſius linee date per fecūdam petitioné circulū.c.b.d.f.

7.4 | Geometric diagrams marked with letters, from Euclid, *Praeclarissimus Liber Elementorum Euclidis perspicacissimi* (Venice: Ratdolt, 1482), 2v. The same figure labeled with Greek letters appears in Bodleian MS D'Orville, 301, fol. 18r. Courtesy of the Smithsonian National Library.

at least as the story of Zeuxis's bird-fooling grapes and Parrhasius's Zeuxis-fooling curtains indicates,[36] though of course what counts as a realistic image is a cultural construction (Gombrich 1969). In the West, images considered naturalistic have long been used to depict the content of a text, or merely to decorate it, even when the text itself lacks any explicit reference to the accompanying image. Indeed, manuscripts are often valued for these unique decorations. But within the first century of printing, improvements in woodcutting and engraving made it possible to reproduce lifelike images in printed texts and to create new occasions for verbal-visual interactions.

While naturalistic images were not historically used as adjuncts to arguments the way diagrams were, they were and still are also used to represent something other than what they depict. They became carriers of further meanings in several historical traditions of linking images to external content. First is the tradition of using elaborately constructed mental images as memory aids in the rhetorical tradition. Rhetorical mnemonics requires the construction of an imaginary architectural space that is then populated with objects serving as visual signs of the material to be remembered. The "pictorial alphabet" suggested by memory images, building on conventional associations between objects and their meanings, is also present in the tradition of *emblems*, illustrations of proverbs and maxims that were popular from the sixteenth century on. An emblem book typically features a woodcut illustrating a biblical or proverbial saying (see examples in Elkins [2001, 195–212]). The image itself sometimes carried a banner with the key verbal text, but it was left to the viewer to decipher the image in terms of the text. A typical emblem image might feature an animal associated with a character type, such as a fox or snake. While naturalistic memory images and emblemata may seem a long way from geometrical diagrams, they nevertheless promote the role of images as bearers of an abstract meaning beyond their value as naturalistic depictions. These prior verbal meanings, assigned by convention to the image, perpetuate the expectation that naturalistic images carry deeper significance. Furthermore, these regimes of imposed meaning require attention to the parts of the image: it is the ram's testicle hanging on the hand of the man in the memory image described in the *Rhetorica ad Herennium* that reminds the legal orator to talk about witness testimony (Cicero 1954, 215). This notion of the meaningfulness of the parts and details of images will persist when images are used in what we would now call scientific texts in the sixteenth and seventeenth centuries.

While it is convenient to separate two traditions of naturalistic and geometric depiction, there have always been important crossovers or even blendings of these two styles of image creation, as arguably occurred in the case of Lister's scallop. To focus on the imposition of the diagrammatic on the naturalistic, art historians have noted the tendency of images not only to be geometrized when they are first drawn but also to become more geometrical from version to version. Parts become more like geometrical figures and are made more symmetrical in relation to one another. This process can be seen in images of the same plant as they appear in consecutive illustrated manuscripts of Dioscorides's herbal *De materia medica*. In the famous early sixth-century CE Vienna Codex, the illustration of a *Juncus maritimus* depicts a naturalistic grassy plant. In a late sixth-century manuscript (which may have been based on the earlier or on another now missing text), the copyist has drawn the thin leaves of the *Juncus maritimus* as orderly intersecting curves, almost as if they were inscribed with a geometer's compass (see Janick and Stolarczyk 2012, fig. 8; for an even more geometrized image in a ninth-century CE copy, see Janick, Whipkey, and Stolarczyk 2013, fig. 4e). William Ivins described such changes in the direction of increased symmetry and regularity as a "rationalization" of these images, and he noted that this "degradation and distortion" can occur not only from drawing to drawing, but also from woodcut to woodcut (1969, 40–42).

On the reverse imposition of the naturalistic on the geometrical, it is worth remembering that the geometrical figures are themselves abstractions from the natural world, as a fish was a model for the rhombus. And while "geometrizing" natural objects may seem to detract from their accurate depiction, it can also represent an imposed understanding.[37] The early modern sciences of astronomy and optics are primarily geometrized visualizations of natural observations, and similar techniques are on display in practical manuals of surveying and cosmography, as in the frontispiece to Apianus's *Introductio geographica*

7.5 | Frontispiece to Apianus' *Introductio geographica* (Ingolstadt, 1533). Frontispiece. Courtesy of the Bayerische Staatsbibliothek.

(1533), where observers make geometric sense of their surroundings (fig. 7.5). The same combining of natural and diagrammatic visualization can be easily illustrated in "mechanics" or "statics," a science that in the sixteenth century included a redescription in mathematical terms of "simple machines." These products of real-world engineering savvy (the lever, screw, pulley, wedge, wheel and axle, etc.) were depicted in naturalistic images of machinery in works such as George Agricola's *De re metallica* (1556), a study of sixteenth-century mining and metallurgy that is well known for its innovative use of technical drawing conventions such as cutaway depictions. Such treatises on machines were among the first printed texts to use letter notations on images of real-world objects. But in studies on the principles behind these machines such as Guido-baldo's *Mechanicorum liber* of 1577, they were depicted first in naturalistic images and then in geometrical diagrams as a tool for analyzing their principles of operation (see figs. 7.6, 7.7; both images are labeled with letters referred to in the text).

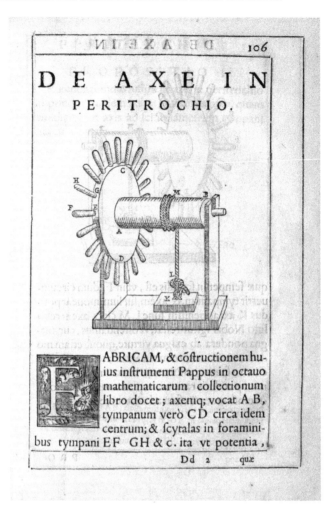

7.6 | Mechanical device with letter notations, from Guidobaldo del Monte, *Mechanorum liber* (Pisa: Hieronymum Concordiam, 1577), fol. 106r. Courtesy of the Max Planck Research Library.

James Franklin described the argumentative uses of such melding when he wrote that "a diagram is a picture in which one is intended to perform [an] inference about the thing pictured, by mentally following around the parts of the diagram" (2000, 55). Naturalistic images can work in the same way as inference-generating diagrams when labels are imposed on the depicted parts.

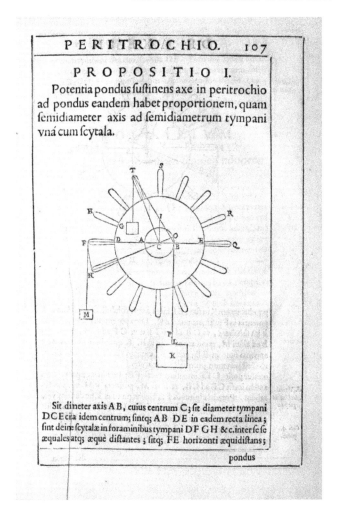

PERITROCHIO.   107

PROPOSITIO I.

Potentia pondus fuſtinens axe in peritrochio
ad pondus eandem habet proportionem, quam
ſemidiameter axis ad ſemidiametrum tympani
vná cum ſcytala.

Sit dimeter axis A B, cuius centrum C; ſit diameter tympani
D C E cſa idem centrum; ſintq; A B D E in eadem recta linea ;
ſint dein: ſcytalæ in foraminibus tympani D F G H &c.inter ſe ſe
æquales atq; æqué diſtantes ; ſitq; F E horizonti æquidiſtans;

pondus

7.7 | Mechanical device with letter notations, from Guidobaldo del Monte,
Mechanorum liber (Pisa: Hieronymum Concordiam, 1577), fol. 107r. Courtesy of
the Max Planck Research Library.

The Pictorial Sciences

The early modern sciences that depended on naturalistic depiction were those
that served medicine, namely anatomy and botany. To begin with anatomy, the
study of any body in dissection is inevitably a visual discipline, and Aristotle
may have originated the practice of using images keyed with letters to indicate

the anatomical parts discussed in the corresponding text (Van Leeuwen 2016, 89–91; the hedging *may* reflect the possibility that the letter references to diagrams are later scribal insertions). In his discussion of cuttlefish embryology in the *History of Animals*, for example, Aristotle writes, "in the case of the young cuttlefish, as in the case of other animals, the eyes at first seem very large. To illustrate this by way of a figure, let A represent the egg, B and C the eyes, and D the young cuttlefish" (Barnes 1:868; 550a 24–26; see also the geometrical arrangement described in the conclusion of *The Movement of Animals*, Barnes 1:1096; 703b 30–35). Presumably a separate set of diagrams accompanied the *History of Animals* with the images that such passages referred to (Leroi 2014, 60). None of these illustrations survive, but Aristotle's descriptive practice of keying text to image with letters (as in geometry texts) offered an authoritative precedent.

While anatomical drawings based on dissections, like the lost illustrations of Aristotle, featured labeled parts, early botanical illustrations did not. The herbals appearing in Germany from Brunfels in 1530 (*Herbarum vivae eicones*) and from Fuchs in 1542 (*De historia stirpium*), noted above, do represent a sudden improvement in the naturalistic depiction of plants *and* the ability to reproduce these naturalistic images in technically superb woodcuts. But they relied on separate verbal descriptions in the adjacent text to highlight the criterial features of the imaged plant with no keying of text to image. Fortunately, visual and verbal presentations of the same plant usually occupied adjacent pages for ease of reference.

It was not until several decades later that the convention appeared of isolating plant parts in separate images, drawn to different scales, and then of labeling those parts with letters or numbers cued to verbal identifications. This practice apparently first appeared in a 1586 herbal, as shown in the image of the pulsatilla (fig. 7.8). This woodcut and others like it were included in *De plantis epitome utilissime* by Joachim Camerarius the Younger, who combined the text of Pietro Mattioli's translation and commentary on Dioscorides with new illustrations, many of which came from Conrad Gesner, who had compiled more than fifteen hundred drawings of plants before his death in 1565 (Arber 1912, 92).[38] Most of Camerarius's printed illustrations look precisely like those in earlier herbals that depicted whole plants. But some offer isolated parts drawn to a different scale, and these images reproduce what Gesner had done in his diagnostic drawings (Ogilvie 2006, 199). In his preface to the reader, Camerarius refers to the fact that in some cases Gesner illustrated parts of a plant ("Benevolo Lectori," [p. 3]),

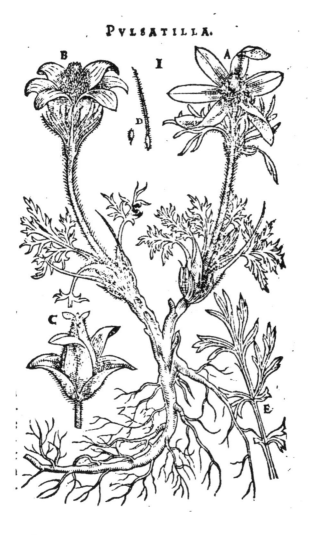

PVLSATILLA.

7.8 | Woodcut of the herb pulsatilla with letter notations, from Pietro Matthioli and Joachim Camerarius, *De plantis epitome utilissima* (Frankfurt: Moenum, 1586), 392. Note: For letter- or number-keyed images not referenced in the adjacent text, see also pp. 27, 103, 173, 174, 178, 217, 261, 276, 361, 468, 500, 509, 51, 516, 523, 571, 609, 619, 680, 699, 727, 736, 774, 783, 791, 807, 811, 845, 858, 871, 891, 916, 948. The numbers and letters often indicate alternate forms, as on p. 656, where atypically the letters are mentioned in the text. Many images bear symbols, some of which may indicate cross sections of stems. See, e.g., pp. 97, 323, 610, 625, 837, 848, 909, 939, 962.

but the identifying letter labels on these parts, faithfully copied from Gesner's originals, do not correspond to anything in the adjacent text. Nevertheless, these images do represent a potentially novel technique and a new approach to identifying plants.

The inspection of plant parts reached a new level of sophistication a century later in work by Marcello Malpighi in the *Anatome plantarum* (1675) and Nehemiah Grew in *The Anatomy of Plants* (1682), works whose titles combine the two pictorial sciences of anatomy and botany. The eighty-three sheets of engravings appended to Grew's work include a series of detailed microscopically viewed quarter-cross-sections of tree branches (tables 22 through 35) that are clearly labeled with letters. Engraved on image after image, the same letters indicate similar structures, allowing comparisons among species, and the book also contains an "Explication of the Tables" giving additional legends for the letter labels. Yet, amazingly, the detailed descriptions of the images in the text, carefully keyed with marginal references to the tables, contain no mention of the letters labeling the parts in the images (see Grew 1682, 108–13). For most of Grew's text, therefore, verbal and visual descriptions remain separate, just as they had for Camerarius a century earlier.

There is, however, a striking exception. At one point, Grew provides a hypothetical causal explanation of how sap might rise in tree trunks, a controversial subject at the time. He speculates that if a vertical vessel carrying sap were surrounded by bladders filled with fluid, the bladders could exert a pressure capable of moving the sap. This explanation is given in the text in terms of the letters labeling a geometrically stylized image (see figs. 7.9 and 7.10). Here, unlike the practice in his other images, a botanical structure is visualized as a rectangular solid (somewhat at odds with the content) and is keyed to an argument based on how the action of parts can sum to an overall effect. The text itself uses the linguistic register of a geometrical proof, though it does not actually offer a proof.

In another early work sponsored, like Grew's, by the Royal Society but much more famous, the language of geometry is also brought in as a way of generating descriptive claims about images. Robert Hooke's *Micrographia* (1665) is well known as a collection of engravings depicting the objects Hooke inspected under his "magnifying glasses." Hooke's images represent "observables" from the real world, though they require a new instrument to be seen, and in the process of depicting them, Hooke the artist broke new ground, as in presenting his images in the circular form of his viewing frame. As far as the text is concerned, Hooke has a great deal to say about his microscope, the laws of refraction, the

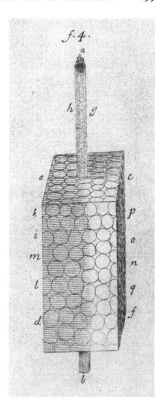

7.9 | Geometrically stylized image for explaining the flow of sap, from Nicholas Grew, *The Anatomy of Plants* (London: W. Rawlins, 1682), table 39. Courtesy of the Library of Congress.

15. §. Let A B be the *Veſſel* of a *Plant*. Let C E D F be the *Bladders* of the *Parenchyma*, wherewith, as with ſo many little *Ciſterns*, it is ſurrounded. I ſay then, that the *Sap*, in the *Pipe* B A, would, of it ſelf, riſe but a few Inches ; as ſuppoſe, from D to L. But the *Bladders* D P, which ſurround it, being ſwelled up and turgid with *Sap*, do hereby preſs upon it ; and ſo not only a little contract its bore, but alſo transfuſe or ſtrain ſome *Portion* of their *Sap* thereinto : by both which means, the *Sap* will be forced to riſe higher therein. And the ſaid *Pipe* or *Veſſel* being all along ſurrounded by the like *Bladders* ; the *Sap* therein, is ſtill forced higher and higher : the *Bladders* of the *Parenchyma* being, as is ſaid, ſo many *Ciſterns* of *Liquor*, which transfuſe their repeated Supplies throughout the length of the *Pipe*. So that by the ſupply and preſſure of the *Ciſterns* or *Bladders* F D, the the *Sap* riſeth to L ; by the *Bladders* Q L, it riſes to M ; by the *Bladders* N M, it riſes to I ; by the *Bladders* O I, it riſes to K ; by the *Bladders* P K, it riſes to E ; and ſo to the top of the *Tree*. And thus far of the *Motion* of the *Sap*.

7.10 | Text accompanying the stylized image explaining the flow of sap, from Nicholas Grew, *The Anatomy of Plants* (London: W. Rawlins, 1682), 126. Courtesy of the Library of Congress.

goals of mechanical philosophy, and much else in his lengthy preface. But when it comes to the images he has a problem, because in fact pictures do not speak for themselves. Hooke has to generate text about his images when there is nothing to say about them other than what they look like. His solution is one that any well-educated gentleman in the seventeenth century would use whose intent was the serious understanding of an object under the eyes: look at it geometrically.

Though he disavows any "infallible Deductions or certainty of Axioms" (Hooke 1665, b1), Hooke begins with the conceit of a deliberate imitation of Euclid: "As in Geometry, the most natural way of beginning is from a Mathematical *point*; so is the same method in Observations and *Natural History* the most genuine, simple, and instructive" (1665, 1). Playing off opening definitions from the *Elements*—*A point is that which has no part. A line is breadthless length. A surface is that which has length and breadth only*—Hooke's first three images present, first, the *points* of a needle and a printed period (fig. 7.11); second, the *line* of a razor's edge; and third, the *plane* of a piece of fine linen.

The text references letters on each image. An "A" marks a circled printer's period. He then describes the top "of a small and very sharp Needle, whose point *a a* nevertheless appeared . . . *irregular* and *uneven*"; "A, B, C [the first two letters barely visible against the dark background] seem'd *holes* made by some small specks of *Rust*; and D some *adventitious body*, that stuck very close to it." Hooke's purpose in citing these labeled imperfections is to contrast the "rudeness and

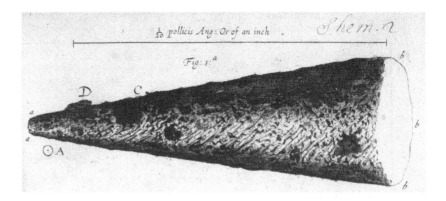

7.11 | Point of a needle and a period, from Robert Hooke, *Micrographia* (London: Martyn and Allestry, 1665), fig. 1 (facing p. 5). Courtsey of Hunt Institute for Botanical Documentation, Carnegie Mellon University, Pittsburgh, PA.

bungling" of human art with the works of God, the seeds, plant parts, and insects on view in many of his later images (1665, 2).

After these opening illustrations and lengthy discussions of experiments with glass tubes, making lead shot, optics, colors, and much more (1665, 10–78), Hooke resumes his catalogue of images, beginning with geometrical forms observed under the microscope in fragments of flint and crystals of frozen urine. Many of these further engravings bear tiny letters connecting parts of the image to details of the verbal descriptions. The famous illustration of the cross-section of cork in *Micrographia*, for which Hooke receives credit as the discoverer of cells, contains only the designation A for the horizontal and B for the longitudinal section (1665, 114). Judged against the standard of dialectical partition, few integral parts are identified in Hooke's images. But keyed to the text with letters, they represent a gesture at control serving a minimal "this exists" argument. The only meaning that can be given to the labeled parts is the fact that they exist as parts that can be labeled. For Hooke, as for Grew and Lister, the conventions for labeling and referencing naturalistic images as though they were geometrical diagrams are firmly in place. They represent an expectation of extra significance from traditional ways of viewing images, but as yet there are no extra meanings to assign.

A little over a hundred years later, however, the labeled parts of naturalistic images would represent a new level of arguing, combining part-whole reasoning with arguments from proportion. This new standard appears in the work of the comparative anatomist Georges Cuvier. Between 1796 and his death in 1832, Cuvier would establish species distinctions and the reality of extinction based on his careful comparative analysis of the skeletons of existing and fossil quadrupeds. He argued convincingly, for example, that the Indian and African elephants were separate species and that the mammoth remains of Eurasia and the mastodon remains of America were not the bones of contemporary animals, as had been assumed, but belonged to distinct species no longer alive. His many publications in comparative anatomy were illustrated by engravings made from his own careful drawings. In some ways, Cuvier's illustrations simply continue the centuries-old tradition of labeling the parts of a dissection, but the arguments for reconstructions supported by his illustrations are triumphs of part/whole reasoning along with elicited ratios and proportions. Indeed, Cuvier would add to Melanchthon's inference rules the possibility of inferring a whole from just one part in order to "recognize a genus or to distinguish a species by a single fragment of bone" (Cuvier 1834, 1:97; see also 1:178; translation mine).

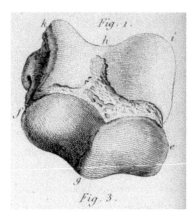
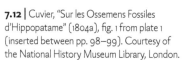

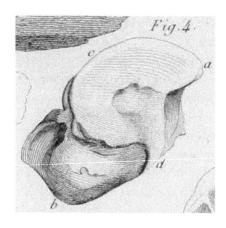

7.12 | Cuvier, "Sur les Ossemens Fossiles d'Hippopatame" (1804a), fig. 1 from plate 1 (inserted between pp. 98–99). Courtesy of the National History Museum Library, London.

7.13 | Cuvier, "Sur les Ossemens Fossiles d'Hippopatame" (1804a), fig. 4 from plate 1 (inserted between pp. 98–99). Courtesy of the National History Museum Library, London.

Cuvier's techniques are on display in his papers on hippopotamus bones, recent and fossil. As usual, he first published on the remains of contemporary specimens, establishing proportions among the parts of a skeleton, before tackling the fossil fragments. His argument on the hippopotamus was then based on careful comparative analysis between the contemporary animal and the fossil bones at hand, but his brilliance was in deciding what to measure in order to create ratios for comparison, as indicated by the letters on his images (see figs. 7.12 and 7.13).[39] At one point, he directs readers to a plate with two figures (1 and 4) giving a frontal and side view of an ankle bone. The text cued to these figures offers not merely descriptions but precise measurements.

Length of the external face, from *a* to *b* 0.117
Vertical height, from *c* to *d* 0.072
Width of tarsal sheave [*poulie tarsienne*], from *e* to *f* 0.107
Distance from the base of the tibial sheath
    to the extremity of the intermediate stopping point
    of the tarsal sheath, from *h* to *g* 0.117
Width of its tibial sheath, from *i* to *k* 0.097
(Cuvier 1804b, 109)

By comparing these measurements with those from a contemporary animal, Cuvier concluded that one of his specimens represented a now extinct species. He was quite aware that the rigor of these identifications rested on mathematical relationships: "In a word," he wrote in a now famous passage, "the form of a tooth determines [*entraine*] the form of the condyle, that [of the condyle] the shoulder blade, that [of the shoulder blade] the claws, just as the equation of a curve determines all its properties" (1834, 1:181). And in one of his most famous demonstrations showing the validity of his techniques, Cuvier inferred from the exposed part of a fossil that its hidden bones were characteristic of a marsupial; he then chipped away the stone to reveal the predicted parts that defined the genus of the whole animal. At the end of the article on this work, Cuvier speculated that as in recent chemistry, so in anatomy, there was no science that could not with the proper methods become almost *geometrique* (Cuvier 1804a, 292).[40]

Conclusion

Students in the sixteenth and seventeenth centuries were taught the discourse arts of rhetoric and dialectic as well as some arithmetic and geometry in Latin schools and universities, and the reformed dialectic of the sixteenth century continued to be a curricular anchor in Protestant universities well into the eighteenth century. Would-be natural philosophers learned, among other methods of arguing, the importance of creating an inventory of parts integral to a whole and of establishing ratios among those parts and proportional relationships between comparable parts across species. The art of dialectic, now a lost art, went further in applying the methods and standards of arithmetic and geometry to any subject that could be argued from universal experience or innate principles. In a text like Melanchthon's *Erotemata dialectices*, the further inclusion of topics and techniques from mathematics, and even of math problems as examples, makes his mid-sixteenth-century version of dialectic a truly general art of reasoning.

While Melanchthon's dialectic foregrounded mathematical standards and methods of reasoning, it also gave a new importance to the power of images in arguing. And in an era when images could be reproduced accurately in woodcuts and engravings for the first time, there were expanded opportunities for arguing from visuals, drawing on the two imaging traditions surviving from antiquity: geometrical diagrams and naturalistic drawings. One tradition represents the manipulation of abstract shapes for proofs, often with practical applications, and the other the depiction of objects in the world, often as carriers of added meanings. In the early modern period, these two traditions merge as the conventions of diagramming are imposed on naturalistic images, allowing easier reference to these images and their parts in texts. Geometric ways of seeing and reasoning are transferred as well, imposing abstract viewing schemas on naturalistic drawings and making the language of geometry available in the accompanying verbal descriptions and arguments.

The processes of arguing with and from geometrically controlled images, common in early modern natural philosophy, also show the increasing imposition of mathematical ways of viewing on the natural world. A commonplace in the history of science, this increasing mathematization is usually equated with the astronomy, mechanics, and physics of Kepler, Galileo, or Newton. But it is also on display in the work of the anatomists and botanists of the sixteenth and seventeenth centuries, who disciplined their scrutiny of natural objects in printed images and verbal descriptions, as the scallop was managed by Martin and Anne Lister. By the time of Cuvier, identifying criterial proportions among identified parts and comparing parts across species yielded even further results in the mathematical interpretation of natural objects. In the reverse process used in the practical arts of architecture and engineering, parts of the world are not interpreted; they are designed and built according to mathematical models. Things intentionally built or planned are of course more easily controlled than things given in nature, and the mathematical management of the material world has only continued and accelerated since the early nineteenth century, until now much of the visual world encountered every day is a mathematically constructed world.

Notes

1. Lister and other seventeenth-century natural philosophers discussed below would have studied one of the many texts derived from Melanchthon's and from similar sixteenth-century dialectics, such as Robert Sanderson's *Logicae artis compendium*, which were in print throughout the seventeenth century in England.

2. Melanchthon also wrote prefaces for a number of other important astronomical works such as Johannes Schöner's *Tabulae astronomicae* 1536. Scholars offer varying lists of these works beginning with Bernhardt 1865, 22–26; K. Reich 1993, 110–12; and U. Reich 2017, 562–70.

3. "Sunt igitur alae mentis humanae, Arithmetica et Geometria." The entire oration is edited by Sachiko Kusukawa and translated by Christine F. Salazar in Melanchthon 1999, 93. It is likely that Melanchthon collaborated on this oration with Rheticus.

4. "Tanta est Dialecticae & Rhetoricae cognatio, vix ut discrimen deprehendi possit. Quidam enim Inventionem ac Dispositionem communem utrique arti putant esse. Ideoque in Dialecticis tradi locos inveniendorum argumentorum, quibus Rhetores etiam uti solent."

5. "Circa omnes materias seu quaestiones, de quibus docendi sunt homines, sicut Arithmetica versatur circa omnes res numerandas. Et magna cognatio est Dialectices & Arithmetices. Deus indidit naturae intelligenti noticias numerorum, ut res discernat. . . . Postquam autem Arithmetica numeravit res, accedit Dialectica, & distinctis rebus nomina diversa & definitiones attribuit, quaerit membra, partes, causas, effectus, & alia, quae accedunt. Deinde in ratiocinando componit cohaerentia, & distrahit diversa" (1548, 1v–2r). Melanchthon's discussion of the affinities between arithmetic and dialectic continues with a comparison between dialectic's procedures and arithmetical reasoning with proportions: if one pound of wax costs one drachma, four pounds will cost four (1548, 2r). See also the definition of arts such as arithmetic as divinely revealed for their usefulness (1548, 22v). Most quotations from the *Erotemata dialectices* are taken from the corrected second, 1548 edition, unless otherwise indicated.

6. He also allies dialectic with geometry when he repeats the traditional stylistic distinction between the spareness of dialectical discourse and the copiousness of rhetoric: "A geometer, for example, would be inept and ridiculous if, like a declaimer, he wanted to add rhetorical devices to his demonstrations." *Geometer erit ineptus & ridiculus, si, ut Declamator, suis demonstrationibus addere volet ornamenta oratoria* (1548, 3r). Melanchthon goes on to say that when people are being taught, the propriety of dialectic is preferable, but on moral matters, where exhortation and fixing one idea are goals, the illuminations of rhetoric are necessary.

7. "Tradidit autem Deus Arithmeticen, Geometriam, physica quaedam & ethica, ut gubernent multas vitae partes, & sint adminicula in doctrina coelesti, ac vult harum doctrinarum firmam & immotam esse certitudinem" (Melanchthon 1548, 131v).

8. The five predicables have to do with the semantic roles for predicates involved in definition. They were understood from antiquity through the Renaissance in terms of Porphyry's commentary on Aristotle's *Categories*, the *Isagoge*. Melanchthon's ten predicaments are Substance, Quantity, Quality, Relation, Activity, Passivity, Time, Place, Situation, and Condition.

9. "Linea est longitudo sine latitudine & profunditate, cuius extrema sunt duo puncta. Sciant autem iuniores, non dici hic de linea picta, sed mente cogitanda est sola longitudo abstracta a corporibus, ut, Mathematicus considerat quantitates sine materia."

10. Traditionally the category of quantity included speech units based on differences in syllable length, an ancient inclusion that Melanchthon actually finds unprofitable. In a subsection titled "Why is speech included in this predicament?" he explains that he wants to move speech sounds into the predicament of quality (1548, 19v).

11. Melanchthon's separation of discrete from continuous quantities reflects a time, as historians of mathematics have pointed out, when the understanding of geometrical magnitude was not really a matter of measurable quantities expressible in numbers since "discrete" numbers were thought of as only positive, whole integers while many geometrical magnitudes (e.g., the hypotenuse of many right triangles) obviously cannot be expressed as integers (Bos 2001, 128–31). Numbers indicating measured lengths do appear in geometrical diagrams at the time, as in Recorde's 1551 practical introduction to the art, but they are always whole numbers. See Barany 2010, 136.

12. "Quarter quinque sunt viginti, Ergo & quinquies quarter sunt viginti. Hic vides conversionis exemplum. Ita in Geometria multa iudicantur conversione" (1548, 67r). Melanchthon

goes on to make the point, cited above, that Aristotle constructed much of dialectic from geometry.

13. From the preface to book VII of the *Synagoge*.

14. Contemporary logicians would require the quantifier *some* in the conclusion.

15. "Aliqui rationem consequentiae sumpserunt ex hoc principio geometrico: Quae sunt aequalia uni tertio, inter se aequalia. Haec sententia cum sit natura nota, confirmat & consequentuam, de qua loquitur. Est autem similis consequentia: Huic docdranti pedis est aequalis haec linea, / Et eidem dodranti est aequalis altera linea, / Ergo hae omnes inter sese sunt aequales. Exempla geometrica sunt illustria. Cumque in his valere consequentiam manifestum sit, valebit & in aliis materiis. Ita propter similitudinem accommodatur huc principium istud geometricum."

16. It is not clear where Melanchthon takes this phrasing from, since the first printed Greek text of Euclid, published by his friend Simon Grynaeus, has a different phrasing for 3.24: "Ta epi ison eutheion omoia tenmata kuklon isa allelois estin" (Grynaeus 1533, 39).

17. "In hoc genere alia magis probabilia & illustriora sunt, alia minus. Cum autem non sunt immotae demonstrationes, ut sunt Geometricae, dicuntur dialectica argumenta. Sic enim vetustas loquebatur, Geometris tribuebat demonstrationes, caetera in materiis physicis & moralibus magna ex parte vocabat dialectica argumenta. Etsi enim in his aliquae sunt demonstrationes, tamen multo maior pars tantum probabilies materias continent."

18. "Et quia inter primos syllogismos sunt, syllogismi Arithmetici, in his perspicuitatem consequentiae consideremus. Natura notae sunt proportiones. In harum collatione, vero syllogismi ordine positis tribus numeris, quartam sequi necesse est, Una libra cerae valet drachma una, / Hic sunt librae decem, / Ergo decem drachmae erunt precium. Hic ordo est primae figura syllogismorum, si dextre accommodes. Nam maior pro universali ponitur, Quaelibet libra valet una drachma, vel quot sunt librae, tot sunt drachmae. Ut autem proportio, quae naturali luce mentis iudicari potest, ostendit consequentiam in talibus syllogismis arithmeticis valere, & immotam esse, Ita in caeteris syllogismis valere consequentiam, & immotam esse conspicitur ex demonstrationibus non procul accersitis, sed obviis & evidentibus."

19. "Speculabilia vocantur noticiae naturales, seu propositiones naturae notae, ex quibus physicae ac mathematicae doctrina extruuntur, ut totum est maius qualibet sua parte. Quae sunt uni aequalia, inter sese sunt aequalia, unius corporis simplicis, unicus est motus naturalis."

20. "Ut autem principia arithmetica & geometrica, & demonstrationes inde factae, firmae & immotae sunt, ita sciamus, revera principia practica & deomstrationes inde factas, firmas & immotas esse. Ut adsensus firmus & perpetuus est, quo adsentior huic propositioni, bis 4 sunt 8, ita firmo adsensu amplecti homines oportebat hanc propositionem: Non faciendum est furtim. Sed in hac caligine minus illam divinam lucem aspicimis, & pravi affectus languefaciunt adsensionem in practicis materiis."

21. "Id est non loquitur de Geometricis aut Arithmeticis principiis & demonstrationibus, ... Interea noticiae in mente suum locum habent, & manent easdem, bis 4 sunt 8. Vita honesta et tranquilla est res bona. Mors est res mala &c." See also the refutation of misinterpretations of Colossians 2:8; in his apparent warning against philosophy, Saint Paul did not, according to Melanchthon, have in mind the divinely revealed truths of arithmetic (1548, 205r).

22. The exception among dialectical textbooks is Rudolph Agricola's influential *De inventione Dialectica*, written ca. 1479 and first printed in 1516. This text drops all discussion of the types of propositions and the forms of the syllogism and concentrates exclusively on the inventional topics closest to the system found in Cicero and Boethius. Ramus followed

Agricola in this truncation. Most sixteenth-century dialectics, like Melanchthon's, keep the rhetorical focus of Agricola but represent the full Aristotelian *Organon*.

23. In book III, Melanchthon had also noted that arguments using reductio per impossibilem were useful in geometry and natural philosophy (1548, 90v).

24. "Sed partes integrales a quantitate seu numero appellationem habent, ut cum molem partimur, seu numeros dividimus, ut trunco sex pedum dissecto in partes bipedales, hae dicuntur partes integrales."

25. A topic labeled "Proportion" appears in earlier dialectics such as Caesarius's 1539 *Dialectica*. Caesarius, however, is concerned with analogy in general, citing examples such as "the eye is to the head as the mind is to the soul" and "as trespass is to trespass so punishment should be to punishment" (1539, 212). There is no discussion of both ratio and proportion and there are no examples involving quantities as in Melanchthon's "new" topic of proportion. Melanchthon is aware of the usual treatment of proportion as analogy: "Moreover the doctrine of proportions does carry over to the comparisons of persons that are discussed in common dialectical textbooks. Thus the captain of a ship is to the ship as a Bishop is to a Church, etc." (Transfertur autem proportionum doctrina & ad personarum collationes, de his loquuntur vulgares libelli dialectici. Sicut se habet gubernator navis ad navim, sic se habet Episcopus ad Ecclesiam & c. [1548, 165r]).

26. "Valde mihi placet mentionem fieri proportionum inter locos inventionis. Postquam enim de substantia, & et de causis dictum est, de quantitate & de numeris, & numerorum collatione etiam saepe dici necesse est, ut in quotidiana vita sumptuum mentio & usitata & necessaria est. Videat paterfamilias, imo videant omnes amantes iusticiae, ne suptuum magnitudo superet facultates. Haec collatio sumitur ex doctrina proportionum."

27. Units of measurement differed from jurisdiction to jurisdiction in sixteenth-century Germany. The Leipzig *eimer* has been estimated at 75 liters, so 6 eimer would equal 450 liters. See https://www.nuffield.ox.ac.uk/users/murphy/measures/files/Table_6_05_08.pdf.

28. "Cum in Cana Galileae Christus sex hydrias aqua impleri iusserit, quam convertit in vinum, si quaeratur, quantum vini donarit sponso iuxta nostras mensuras, id ostendet Analogia. Una Hydria continebat duas metretas, Metreta vero continebat triginta mensuras, quas nostra appellatione vocamus kannen, Ergo duae metretae continent mensuram, quam nun usitate vocamus ein Eimer. Iam sicut se habet una hydria ad duas metretas, ita se habent sex hydriae ad duodecim metretas. Donavit igitur mensuras vini, ut nos vocamus, sex eimer, quod fuit munus non exiguum."

29. Definition 5 offers the rationale for proportionality: "Magnitudes are said to be in the same ratio, the first to the second and the third to the fourth, when, if any equimultiples whatever are taken of the first and third, and any equimultiples whatever of the second and fourth, the former equimultiples alike exceed, alike are equal to, or alike fall short of the latter equimultiples respectively taken in corresponding order" (Heath 1952; translation). Definition 5 was often misunderstood due to textual difficulties in the transmission of Euclid.

30. The diagonals indicate contradictions between statements (*all are* versus *some are not* and *all are not* versus *some are*), the line connecting the top left and right corners signals contraries (*all are* versus *all are not*—both cannot be true), the line connecting the bottom left and right corners indicates subcontraries (*some are* versus *some are not*—both can be true), and finally the lines joining top and bottom left and top and bottom right statements connect them as subalternates, the top implicating the bottom, *all are* includes *some are* and *all are not* includes *some are not*. All of these relationships are inferences. For an example of such a square fully labeled, see Melanchthon (1555, 125).

31. I am indebted to my colleague Professor Vessela Valiavitcharska for bringing this visualization practice to my attention. See also Panizza 1999.

32. While the combined crescent-shaped figure does not represent a construction in Euclid, such a figure does appear in the Scholion to a ninth-century MS of Euclid (Bodleian MS D'Orville 301 fols. 118v–119r).

33. It is not certain that this addition and other changes in the 1560 edition are attributable to Melanchthon, who died in April of that year, or whether they were made by his son-in-law Caspar Peucer.

34. "Definitio rei est oratio, quae essentiam, aut causas, aut partes, aut accidentia rei exponit. Et quid intersit inter definitionem nominis & definitionem rei, sic intelligi facile potest, Definitio nominis est, cum peregrinae linguae vocabulum interpretaris notiore vocabulo nostrae linguae, & genus nominas, ut si dicas: Centaurium est herba, quam vocamus tausent gulden, vel aurin, genus & nomen audis minus peregrinum, ac fieri potest, ut res adhuc ignota sit. Sed si herba proferatur, ut subiectam oculis intueri possis, iam habes rei definitionem illustrem. Vetus enim dictum est, & dignum memoria: Omnis intuitiva noticia est definition."

35. They include a papyrus found at Herculaneum and dated to the first century BCE (Manders 2008, 97n15) and another text plus image fragment preserved in the dry sands of Oxyrhynchus containing a proof of Euclid 2.5 and dating between 75 and 125 CE. https://www .maa.org/press/periodicals/convergence/mathematical-treasure-euclid-proposition-on -papyrus. Much older geometrical images associated with surveying records have been found, including a Babylonian tablet from 1700 BCE (Heilbron 1998, 32).

36. Notions of "realistic" depiction are further complicated by the ability of artists to create images of things that never existed in nature, like the harpies, minotaurs, and fauns imaged on early Greek vases. These "imagined" depictions tend to be decipherable as composites of the observable: the minotaur, for example, has the head of a bull on the body of a man. But when the medium itself cannot speak to its own value as a record of what was observed rather than imagined (as for instance in "nature prints" versus drawings), it is difficult to know the status supposed for the object depicted on the basis of the image alone.

37. The use of diagrams and naturalistic images in astronomy, optics, and mechanics has been investigated by historians of science, most notably in the recent project on "Diagrams, Figures and the Transformation of Astronomy, 1450–1650" carried out at the University of Cambridge. See Fay and Jardine 2013.

38. Coincidentally, Melanchthon had dedicated the *Erotemata dialectices* to the author of this 1586 medical herbal, Camerarius the Younger, the son, then fourteen years old, of his best friend, Joachim Camerarius.

39. Many of Cuvier's illustrations were not published with letter keys. Indeed, Cuvier thought that in many cases a mere glance at compared illustrations would make an obvious case for differences, as, for example, in the juxtaposed images of Indian and African elephant skulls as proof of their belonging to different species (Rudwick 1976, 20).

40. "Il n'est aucune science qui ne puisse devenir presque géométrique; les chimistes l'ont prouvé dan ces derniers temps pour le leur; et j'espère que le temps n'est pas éloigné où l'on en dira autant des anatomistes."

References

Agricola, Georgius. 1556. *De re metallica libri XII*. Basel: Froben.
Apianus, Petrus. 1533. *Introductio geographia*. [Ingolstadt: Apianus.]
Arber, Agnes. 1912. *Herbals, Their Origin and Evolution: A Chapter in the History of Botany*. Cambridge: Cambridge University Press.

Barany, Michael J. 2010. "Translating Euclid's Diagrams into English, 1551–1571." In *Philosophical Aspects of Symbolic Reasoning in Early Modern Mathematics*, edited by A. Heeffer and M. Van Dyck, 125–63. London: College Publications.

Barnes, Jonathan, ed. 1984. *The Complete Works of Aristotle*. The Revised Oxford Translation. 2 vols. Princeton: Princeton University Press.

Bernhardt, Wilhelm. 1865. *Philipp Melanchthon als Mathematiker und Physiker*. Wittenberg: Vereins für Heimatkunde.

Billingsley, Henry. 1570. *The Elements of Geometrie of the Most Auncient Philosopher Euclide of Megara*. London: Daye.

Bos, Henk J. M. 2001. *Redefining Geometrical Exactness: Descartes' Transformation of the Early Modern Concept of Construction*. Amsterdam: Springer.

Caesarius, Johannes. 1539. *Dialectica cui adjecimus Isagogen in decem Aristotelis praedicamenta*. Cologne: [n.p.].

[Cicero]. 1954. *Rhetorica ad Herennium*. Translated by Harry Caplan. Cambridge, MA: Harvard University Press.

Cifoletti, Giovanna. 2006. "Mathematics and Rhetoric: Introduction." *Early Science and Medicine* 11 (4): 369–89.

Cuvier, Georges. 1804a. "Mémoire sur le squelette presque entier d'un petit quadrupède du genre des sarigues trouvé dans la pierre à plâtre des environs de Paris." *Annales du Museum National d'Histoire Naturelles* 5:277–92.

———. 1804b. "Sur les ossemens fossiles d'hippopotame." *Annales du Museum National d'Histoire Naturelles* 5:99–122.

———. 1834. *Recherches sur les Ossemens Fossiles*. Vol. 1. Paris: D'Ocagne.

Del Monte, Guidobaldo. 1577. *Mechanorum liber*. Pisa: Hieronymum Concordiam.

De Risi, Vincenzo. 2016. "The Development of Euclidean Axiomatics: The Systems of Principles and the Foundations of Mathematics in the Editions of the *Elements* in the Early Modern Age." *Archive for History of Exact Sciences* 70:591–676.

Eck, Johannes. 1517. *Elementarius dialectice*. Augsburg: Officina Millerana.

Elkins, James. 1999. *The Domain of Images*. Ithaca: Cornell University Press.

Euclid. 888. *Elementa, Books I–XV*. Digital Bodleian: MS D'Orville 301. https://digital.bodleian.ox.ac.uk/inquire/p/d1938266-7a9e-4be1-ba0d-af83615d9e1f.

———. 1482. *Praeclarissimus liber elementorum Euclidis in artem geometriae*. Venice: Erhard Ratdolt.

Fay, Isla, and Nicholas Jardine. 2013. "Introduction: New Light on Visual Forms in the Early-Modern Arts and Sciences." *Early Science and Medicine* 18 (1–2): 1–8.

Franklin, James. 2000. "Diagrammatic Reasoning and Modelling in the Imagination." In *1543 and All That*, edited by Guy Freeland and Anthony Corones, 53–116. Dordrecht: Kluwer.

Gombrich, E. H. 1969. *Art and Illusion: A Study in the Psychology of Pictorial Representation*. Princeton: Princeton University Press.

Grew, Nehemiah. 1682. *The Anatomy of Plants*. London: Rawlins.

Grynaeus, Simon, ed. 1533. *Stoicheion Bibl. XV ek ton Theobos Synoysion. Proklou Bibl. IV.* Basel: Herwagen.

Heath, Sir Thomas L. 1952. *Great Books of the Western World*. Vol. 11, *The Thirteen Books of Euclid's Elements*. Chicago: Encyclopedia Britannica.

Heeffer, Albrecht. 2010. "From the Second Unknown to the Symbolic Equation." In *Philosophical Aspects of Symbolic Reasoning in Early Modern Mathematics*, edited by A. Heeffer and M. Van Dyck, 57–101. London: College Publications.

Heilbron, J. L. 1998. *Geometry Civilized: History, Culture, and Technique*. Oxford: Clarendon Press.

Hooke, Robert. 1665. *Micrographia: Or, Some Physiological Descriptions of Minute Bodies Made by Magnifying Glass.* London: Martyn and Allestry.

Ivins, W. M., Jr. 1969. *Prints and Visual Communication.* New York: Da Capo Press.

Janick, Jules, and John Stolarczyk. 2012. "Ancient Greek Illustrated Dioscoridean Herbals: Origins and Impact of the *Juliana Anicia Codex* and the *Codex Neopolitanus.*" *Notulae Botanicae Horti Agrobotanici* 40 (1): 9–17.

Kuspit, Donald B. 1973. "Melanchthon and Dürer: The Search for the Simple Style." *Journal of Medieval and Renaissance Studies* 3:177–202.

Leeuwen, Joyce van. 2016. *The Aristotelian Mechanics: Text and Diagrams.* Cham, Switzerland: Springer International.

Leroi, Armand Marie. 2014. *The Lagoon: How Aristotle Invented Science.* New York: Viking.

Lister, Martin. 1695. "The Anatomy of the Scallop." *Philosophical Transactions [of the Royal Society]* 19:567–70.

Mack, Peter. 1993. *Renaissance Argument: Valla and Agricola in the Traditions of Rhetoric and Dialectic.* Leiden: Brill.

Manders, Kenneth. 2008. "The Euclidean Diagram." In *The Philosophy of Mathematical Practice*, edited by Paolo Mancosu, 80–133. New York: Oxford University Press.

Matthioli, Petrus. 1586. *De plantis epitome utilissima: Novis iconibus et descriptiones pluribus nunc primum diligenter aucta a D. Ioachimo Camerario.* Frankfurt: Moenum.

Melanchthon, Philip. 1532. *Elementorum rhetorices libri duo.* Hagenau: Setezer.

———. 1547. *Erotemata dialectices: Continentia fere integram artem, ita scripta, ut iuventuti utiliter proponi possit.* Wittenberg: Lufft.

———. 1548. *Erotemata dialectices: Continentia fere integram arte ita scripta ut iuventuti utiliter proponi possint; Haec secunda editio emendatior est priore.* Wittenberg: Lufft.

———. 1555. *Erotemata dialectices: Continentia fere integram arte ita scripta ut iuventuti utiliter proponi possint; Et hoc anno 1555 recognita atque locupletata.* Wittenberg: Crato.

———. 1560. *Erotemata dialectices. Continentia fere integram artem, ita scripta, ut iuventuti utiliter proponi possint [sic].* Wittenberg: Crato.

———. 1963a. *Corpus reformatorum, Philippi Melanchthonis opera, quae supersunt omnia.* Edited by Carl Bretschneider. Vol. 11. New York: Johnson.

———. 1963b [1552/1580]. *Erotemata dialectices: Continentia fere integram artem, ita scripta, ut iuventuti utiliter proponi possint.* In *Corpus reformatorum, Philippi Melanchthonis opera, quae supersunt omnia*, edited by C. Bretschneider, 13:508–759. New York: Johnson.

———. 1999. *Orations in Philosophy and Education.* Edited by Sachiko Kusukawa. Translated by Christine F. Salazar. Cambridge: Cambridge University Press.

Murdoch, John E. 1963. "The Medieval Language of Proportions: Elements of the Interaction with Greek Foundations and the Development of New Mathematical Techniques." In *Scientific Change*, edited by A. C. Crombie, 237–71. New York: Basic Books.

Ogilvie, Brian W. 1912. *The Science of Describing: Natural History in Renaissance Europe.* Chicago: University of Chicago Press.

Panizza, Letizia. 1999. "Learning the Syllogism: Byzantine Visual Aids in Renaissance Italy—Ermolao Barbaro (1454–93) and Others." In *Philosophy in the Sixteenth and Seventeenth Centuries: Conversations with Aristotle*, edited by Constance Blackwell and Sachiko Kusukawa, 22–47. London: Routledge.

Panofsky, Erwin. 1955. *The Life and Art of Albrecht Dürer.* Princeton: Princeton University Press.

Reich, Karin. 1998. "Melanchthon und die Mathematik seiner Zeit." In *Melanchthon und die Wissenschaft seiner Zeit*, edited by Gunther Frank and Stefan Rein, 105–21. Sigmaringen: Thorbecke.

Reich, Ulrich. 2017. "Mathematik." In *Philipp Melanchthon: Der Reformator zwischen Glauben und Wissen; Ein Handbuch*, edited by Günter Frank, 559–76. Berlin: Walter de Gruyter.

Roos, Anna Marie. 2012. "The Art of Science: A 'Rediscovery' of the Lister Copperplates." *Notes and Records of the Royal Society* 66:19–40.

Rudwick, Martin. 1976. "The Emergence of a Visual Language for Geographical Science, 1760–1840." *History of Science* 14:149–95.

Schaar, Eckhard. 1988. "A Newly Discovered Proportional Study by Dürer in Hamburg." *Master Drawings* 36 (1): 59–66.

Stifelius, Michael. 1544. *Arithmetica integra*. Nuremberg: Johan. Petreium.

Unwin, Robert W. 1995. "A Provincial Man of Science at Work: Martin Lister, F. R. S., and His Illustrators, 1670–1683." *Notes and Records of the Royal Society* 49 (2): 209–30.

Vesalius, Pedanus. 1543. *De humani corporis fabrica*. Basel: Oporinus.

Westman, Robert. 1975. "The Wittenberg Interpretation of the Copernican Theory" [with Discussion Record]. In *The Nature of Scientific Discovery*, edited by Owen T. Gingerich, 393–457. Washington, DC: Smithsonian Institution Press.

Part 4

Mathematical Presentations: Experts and Lay Audiences

8

Accommodating Young Women | Addressing the Gender Gap in Mathematics with Female-Centered Epideictic

James Wynn

The STEM fields of science, technology, engineering, and mathematics have for the past three decades been important sites of research for feminist scholars in the rhetoric, history, and sociology of science interested in the ways in which social and institutional biases disenfranchise women and influence the character of knowledge in the technical sphere. Sociologists and historians of science have led the way in these areas of research by investigating the influence of masculine metaphors on scientific epistemology (Keller 1995), the role of social change in the exclusion of women from scientific institutions (Schiebinger 1989; 2008), and the conflicts over the possibility of developing a feminist theory of science (Harding 1986). Rhetorical scholars have expanded these investigations in important ways by focusing on the influence of androcentric language and argumentation as well as feminist efforts to resist it. In *Science on the Home Front: American Women Scientists in World War II*, for example, Jordynn Jack (2009) investigates the masculine biases in the genre conventions of scientific writing, and Wendy Hayden (2013) examines how feminists used scientific language to avoid masculinized Enlightenment perspectives on free love in *Evolutionary Rhetoric: Sex, Science, and Free Love in Nineteenth-Century Feminism*.

Though feminist scholars in rhetoric, sociology, and history have broadly investigated the influence of masculine bias and the efficacy of feminist resistance in the sciences, less attention has been specifically paid to their influence in mathematics. Feminist scholar Suzanne Damarin highlights the lack of attention to the relationship between gender and mathematics in her article "Toward Thinking Feminism and Mathematics Together," in which she writes, "If studies of women, gender, and mathematics conducted by mathematicians, psychologists, and mathematics educators have paid scant attention to advances in feminist and cultural theories, neither have scholars in feminist theory, feminist science and technology studies, or interdisciplinary women's studies paid much attention to mathematics" (2008, 114).

Notable exceptions to the state of scholarship identified by Damarin are Claudia Henrion's *Women in Mathematics: The Addition of Difference* and Margaret Murray's *Women Becoming Mathematicians: Creating a Professional Identity in Post–World War II America*. In *Women in Mathematics*, Henrion highlights the "underlying [masculine] beliefs, attitudes, assumptions, and expectations of the [mathematics] community" (1997, xviii–iv). Using data from interviews with eleven women with successful mathematical careers, she shows how masculine attitudes about rugged individualism and beliefs that mathematics is a "young man's game" work against women who want to pursue careers in mathematics. Using similar ethnographic methods, Margaret Murray examines the struggles of female mathematicians who received their PhDs in the postwar period (1940–59), a time when the numbers of women in mathematics were at a historical low. Drawing on interviews with almost two hundred female mathematical scholars, Murray identifies what she calls the "myth of the mathematical life course" and explores how women mathematicians succeeded under the unsupportive social and cultural conditions created by this myth (2000, xi).

Around the same time that Henrion's and Murray's research appeared, a substantive academic discussion in education and psychology was also underway concerning the relationship between gender stereotyping and the performance and retention of female students in mathematical courses of study. Papers in this conversation focused primarily on the problem of *stereotype threat*, "the psychological threat associated with the awareness that one may be viewed through the lens of a negative stereotype" (Pronin, Steele, and Ross 2004, 152). Ethnographic research by sociologist Elaine Seymour on why female undergraduate students switched majors, for example, described how the social pressures of the male-dominated STEM classroom forced women to alter their expression of femininity to deflect stereotyping (1995, 449–51). Building on this work, psychological researchers Emily Pronin, Claude Steele, and Lee Ross detailed how prolonged exposure to classes with a quantitative focus increased the degree to which female students believed they would be negatively evaluated for some feminine behaviors while other feminine behaviors remained unaffected, resulting in a bifurcation of female identity (2004, 153). Additionally, Steven Spencer, Claude Steele, and Diane Quinn explored the potential influence of stereotype threat on testing outcomes in mathematics, verifying the hypothesis that underperformance by female students in mathematics testing can be explained by anxieties about confirming negative stereotypes of female mathematical abilities (1999, 7).

Collectively, researchers in sociology, history, psychology, and education have made important strides in identifying the social and psychological factors that have kept young women and girls from participating in and contributing to the study of mathematics. This chapter extends the current discussion about mathematics in feminist studies by reversing its traditional path of inquiry. Whereas the majority of research has been dedicated to understanding how masculinized institutional practices and social stereotypes have kept women out of mathematics,[1] I will consider the means of persuasion used to attract them to it. Toward this end, I undertake in this chapter a detailed rhetorical analysis of the language and argument in *Math Doesn't Suck*, the first in a series of mathematical reference books developed by TV star and mathematical scholar Danica McKellar to persuade middle school girls to study and pursue careers in mathematics. Guided by the rhetorical genre of *epideictic*, the chapter identifies the strategies McKellar uses to attract her young female audience to the study of mathematics. It also qualitatively analyzes reviews of the book and public commentary on its strategies to explore whether McKellar successfully connects with her primary audience and how public commentators respond to her female-centered strategies to attract young women.

The Problem of Women in Mathematics

Before analyzing the strategies McKellar uses in *Math Doesn't Suck* to attract middle school girls to mathematics, it is important to consider in some detail the circumstances that compelled her to write the book in the first place. This section explores those circumstances by examining congressional deliberations over female participation in science, technology, engineering, and mathematics (STEM) as well as McKellar's involvement in those deliberations.

The problem of women in STEM has been the topic of congressional discussion and action for decades. It was first formally raised in 1980, when the U.S. Congress deliberated on and eventually passed the Women in Science Equal Opportunity Act, and was revisited in 1988 with the creation of the National Task Force on Women, Minorities, and Persons with Disabilities in Science and Technology. The deliberations provide evidence of the underrepresentation of women in STEM fields and arguments about the negative consequences of their underrepresentation. In his opening statement at the 1980 hearings, Senator Howard Metzenbaum (Dem-OH) expressed concern about the gap between men and women working in STEM professions. He stated, "today, fewer than 3 percent of our nation's engineers, 4 percent of our physicists, and 11 percent of our chemists are women."[2] Later in the same hearing, Betty Vetter, executive director of the Scientific Manpower Commission, pointed to the low number of women pursuing postsecondary degrees in the STEM fields as the cause of low numbers of women in the workforce: "Over this decade women have earned 20 percent of all the Ph.D.'s awarded, but only 9.2 percent in the EMP [engineering, math, and physics] fields, and only 14 percent in all of the science fields."[3] Testimony by participants in consequent deliberations highlighted the significant social and economic problems that would likely result from low levels of female participation in STEM-related fields. In his opening statement at the 1988 House hearings, for example, Doug Walgren, chairman of the Subcommittee on Science, Research, and Technology, argued, "the underrepresentation of these groups [women, minorities, and persons with disabilities] in science and engineering raises important concerns about both equal opportunity and the future ability of our society to produce enough scientists and engineers to enable us to compete in an international economy."[4]

In their assessment of the problem of gender inequality in the STEM disciplines, lawmakers and experts from academia and the private sector identified three root causes of inequality that they believed could be dealt with through targeted action. These causes included the inadequacy of primary and secondary education in the sciences, the inequalities of opportunity for women in higher education and the workforce, and the lack of public awareness among women about careers in the sciences and their benefits. Of these root causes, the first and the third are addressed by McKellar in *Math Doesn't Suck*; therefore, it is useful to examine these two causes and the proposed strategies for addressing them in further detail. In addressing the first root cause, lawmakers divided up the inadequacies of primary and secondary education into two main problems: the *quality* of the education and its *relevance* for students. They collectively agreed that these problems were the consequence of a lack of adequate training of educators and the absence of classroom strategies and materials they could use to promote learning about and interest in science and mathematics. To address these problems, the Women in Science Equal Opportunity Act authorized the National Science Foundation (NSF) to develop "methods and instructional materials and technologies to improve the quality and relevance of education in science and mathematics and to increase student awareness of career opportunities requiring scientific and technical skills."[5] It also authorized the NSF to train and retrain "teachers, counselors, [and] administrators . . . to improve the quality and relevance of education in science and mathematics and to increase student awareness of career opportunities."[6]

In addition to addressing inadequacies in education, lawmakers also targeted problems in the popular perception of science and mathematics that they believed contributed specifically to women deciding not to pursue careers in these fields. They identified two primary problems: (1) a lack of public awareness about the importance of women's participation in STEM fields and (2) the presence of stereotypes about professionals in these fields that might discourage women from pursuing careers in them. To raise public awareness of the careers for women in STEM, Congress proposed that the NSF "improve the scope, relevance, and quality of information available to the public concerning the importance of the participation of women in careers in science and technology through the use of radio, television, journals, newspapers, magazines and other media."[7] To address gendered and racial stereotypes of science as a white, male profession, Congress also tasked the NSF with creating educational materials to neutralize gender stereotypes and highlight the contributions of women to science. Toward these ends, the Women in Science Equal Opportunity Act authorized the NSF "to support the development of books and instructional materials [which] . . . portray women in scientific and technical careers . . . [and] present scientific and technical material in a manner which is not biased on the basis of gender."[8] It also authorized the creation of a visiting women scientists program whose female participants would "visit secondary schools and institutions of higher education in all regions of the country [to] . . . encourage girls and women to acquire skills in mathematics and science . . . [and to] encourage girls and women to consider careers in science and engineering and to prepare themselves appropriately for such careers."[9]

As a result of the efforts by Congress and the NSF in the 1980s to study the lack of female participation in science and address its causes, substantive gains were made in the number of women pursuing degrees and jobs in math and the sciences in the decade that followed. By the late 1990s, for example, women accounted for 33 percent of all the PhDs awarded in engineering and the sciences, more than double the 14 percent reported in 1980 (NSF 2000, xi). They also constituted 23 percent of the total workers in science and engineering jobs, more than doubling the 9.8 percent reported in the early 1980s (NSF 2000, 51). In mathematics, the numbers of women following academic trajectories were more modest but still impressive. Between 1980 and 1996, the number of women earning bachelor's and master's degrees as a percentage of all degrees earned in mathematics rose by approximately 4 percent for each category[10] (NSF 2000, 119, 170). During the same period, however, the number of women as a percentage of all PhDs earned in mathematics jumped dramatically, almost doubling from 12.8 percent in 1980 to 23.4 percent in 1996 (NSF 2000, 180).

Despite these gains in education and employment, women still remained underrepresented in STEM fields, encouraging lawmakers to continue studying and taking action on the problem.[11] In 1998 Congresswoman Constance Morella (Rep-MD) formed the Commission on the Advancement of Women and Minorities in Science, Engineering, and Technology Development (known as the Morella Commission). The purpose of the commission was to conduct meetings and public hearings across the United States with "industry, govern-ment, academe, and the non-profit sector" to understand the obstacles faced by women and minorities in STEM fields and to find "examples and best practices and effective strategies for making [STEM] ... education and careers more acces-sible."[12] The report of the commission's findings—*Land of Plenty: Diversity as America's Competitive Edge in Science, Engineering, and Technology*—became the subject of public hearings in 2000 in which a number of experts testified about possible solutions to the problem. Among these experts was the actress Danica McKellar, a TV celebrity whose mathematical accomplishments and experience with the challenges facing women in mathematics qualified her as an expert on the subject. Examining her participation in these congressional hearings shows her commitment to addressing the problems facing women in mathematics and provides insight into her efforts to attract women to study and pursue careers in mathematics in *Math Doesn't Suck*.

Danica McKellar: Advocate for Women in Mathematics

Danica McKellar is probably most well known for her role in the hit television series *The Wonder Years*, in which she played Winnie Cooper, the intelligent and beautiful classmate and love interest of the show's main character, Kevin Arnold (played by Fred Savage). After the series ended in 1993, McKellar enrolled as a student at UCLA, where she initially intended to study writing and directing. After finding success in her college math classes, however, she decided instead to major in mathematics. McKellar excelled in her mathematical studies and even published original scholarship. In the summer before her senior year (1997), she and another female student, Brandy Winn, were encouraged by their professor, Lincoln Chayes, "to prove a property that would indicate when the magnetic field would line up in a certain direction" (Chang 2005). After months of work, the finished proof, known as the Chayes-McKellar-Winn theorem, was published in the prestigious *Journal of Physics A: Mathematics* under the title "Percolation and Gibbs States Multiplicity for Ferromagnetic Ashkin-Teller Models on Z^2" (Chang 2005).

After graduating from UCLA in 1998 with a degree in mathematics, McKellar decided to return to acting rather than pursue graduate school because she found the academic world "too isolating and lonely" (Chang 2005). She continued, however, to be involved with mathematics by hosting a website where she would "answer math questions in an attempt to be a role model, and . . . to encourage young people . . . especially girls to stay interested in math and to have fun with it."[13] McKellar also became the national spokesperson for Figure This!, "a government campaign designed at getting middle school kids interested in math and doing it with their parents."[14]

As national spokesperson for Figure This!, McKellar was asked to participate in congressional hearings on the Morella Report in the summer of 2000. In her testimony, she explained her views on why she believed girls and young women stay away from mathematical fields and how she thought their perspectives might be changed. In the written portion of her testimony, McKellar identified two main obstacles that she believed impeded girls and young women from pursuing careers in mathematics or in other scientific and technical fields that require mathematical training: (1) their lack of preparation in the subject and (2) their lack of interest. In her estimation, both of these problems arose in middle school, a crucial point in female students' intellectual and social development. It is at this point in their education, she argued, that foundational mathematical concepts and operations that require sophisticated mental operations, such as fractions and solving algebraic equations, are introduced. If students are not able to pick up these building blocks of higher mathematics, their chances of keeping up intellectually and staying engaged in mathematics significantly diminish. She testified: "Before middle school, math class amounts to memorizing the techniques of basic arithmetic. . . . However, beginning in middle school, math entails more than rote memorization of methods. Suddenly, students are asked to swallow concepts like fractions, percentages, negative numbers, etc."[15]

In addition to increased sophistication in the level of mathematics, McKellar also recognizes a significant shift in the social behavior of girls. Their changing hormones precipitate alterations in social self-awareness, and their need to fit in encourages them to adopt communally shared identities for gendered behavior. In this context, mathematics, which is stereotypically connected with white nerdy males, represents the antithesis of the social image that, in McKellar's opinion, young females are trying to cultivate: "While girls are forming their self-image in those pivotal middle-school years, they need to be exposed to positive images of female scientists to combat the stereotypical white male images heavily portrayed in the media. . . . There are many young women that are extremely talented in the sciences but who do not even identify their own talent not believing they would make good scientists because of deeply-rooted social conditioning."[16]

After laying out her opinion on the problems associated with the mathematical education of middle school girls, McKellar proposed two interventions that she believed might effectively address the educational challenges of middle school math. To deal with the problem of student preparation, she proposed that the government spend more money to identify, train, and improve the salaries of middle school math teachers. Supporting the development of a qualified workforce of teachers would enable students to get the professional help they needed to transition into the more complex subjects of mathematics.[17] To solve the second problem of the lack of role models, she suggested introducing female math and science role models into schools, a perennially favored solution to this problem. She also included, however, a less commonly considered suggestion, that the leading national associations and foundations of mathematics and science develop a national PR campaign to combat the white, male, and nerdy image of science:

> We have got to show math and science in their true light to make them exciting and interesting to all students but especially to the underrepresented demographics [like women and minorities]. Let's call on the ambition of young people! What do young people usually aspire to be? Movie stars, sports heroes—glamorous jobs. Why not make SET [Science Engineering and Technology] careers more glamorous for kids? After all . . . the war for talent makes technically skilled workers more and more

valuable so *let's make that known*. We can appeal to the desire kids have to be *special* and *valued*, and steer kids away from the fear of ridicule of being nerdy.[18]

In McKellar's testimony, she suggested that Congress focus their attention on the education of middle school girls and on creating a sense within this demographic of identification with and excitement about careers in mathematics and science. As the analysis of McKellar's text will show, these recommendations presage the strategic rhetorical decisions she makes in *Math Doesn't Suck*. In the text, McKellar provides middle school girls with clear and well-thought-out explanations and examples to help them grasp essential subjects such as factors, fractions, and decimals. More importantly, however, McKellar makes a conscious effort to take on stereotypes about mathematics by strategically choosing content, language, and style that she hopes will persuade her audience to identify mathematics as relevant to their feminine identities. She also offers her young female readers alternative positive framings of mathematics throughout the text that highlight the field's beneficial and glamorous dimensions in hopes of keeping them focused on their mathematics education and encouraging them to consider careers in mathematical fields.

Epideictic Rhetoric

A historical overview of deliberations about how to attract women to STEM and the factors that have kept them away provides evidence of the social need for a math book specifically designed to attract women to mathematics. Though few rhetorical scholars have investigated strategies to attract audiences, female or otherwise, to mathematics,[19] they have examined efforts by scientists and science writers to attract nonexpert, media-consuming publics to writing about science. In "Accommodating Science: The Rhetorical Life of Scientific Facts," for example, Jeanne Fahnestock explores how the content and argument of scientific articles change as scientific accommodators adapt them for nonexpert audiences. One of the important transformations that these pieces make is that they move from a *forensic* genre of rhetoric, in which facts and causes are recounted, to an *epideictic* genre, in which writers praise science to attract their audience's attention to it. Fahnestock explains: "Accommodations of scientific reports . . . are not primarily forensic. With a significant change in rhetorical situation comes a change in genre, and instead of simply reporting facts for a different audience, scientific accommodations are overwhelmingly epideictic; their main purpose is to celebrate rather than validate" (1986, 33). Given epideictic genre's applicability to writing designed to attract nonexperts to scientific topics, it may also prove useful for describing accommodations designed to attract female audiences to mathematics. Testing this hypothesis requires considering what the features of the epideictic genre are and whether these features correlate with characteristics of McKellar's rhetorical situation and the strategies she uses in her text. If these characteristics prove compatible, then McKellar's work can be considered the first classified instance of mathematical epideictic and an important extension of rhetorical theory to the study of mathematics.

For epideictic to be considered an appropriate and useful genre to apply to assessing McKellar's argument, it is important to identify its genre characteristics and consider what they might reveal about her persuasive strategies in *Math Doesn't Suck*. Perhaps the most influential modern description of the epideictic genre appears in Perelman and Olbrechts-Tyteca's *The New Rhetoric*, in which the authors define epideictic as a genre of argument whose goal is to increase a disposition to act by increasing the audience's adherence to values that the speaker believes are already generally accepted by them. They explain, "the adherence gained by a speech can always advantageously be reinforced. It is in this perspective that epideictic oratory has significance and importance for argumentation because it strengthens disposition toward action by increasing adherence to the values it lauds" (1971, 50). In these lines, Perelman and Olbrechts-Tyteca highlight two of the most important features of epideictic argument. First, epideictic is related to action. In contradistinction to deliberative or forensic rhetoric, however, the action that the audience is being persuaded to take is not immediate to the audience's hearing or reading of an argument (like a vote in the Senate or a decision by a jury) but occurs in an alternative future context. They explain, "the sharing of values [in an epideictic speech] is an end pursued independently of the precise circumstances in which the communion will be put to the test" (1971, 53).

A second, important characteristic of the epideictic genre is that it involves argument from values. In particular, the goal of epideictic is to promote future action by increasing the audience's adherence to values the speaker believes they already hold. Perelman and Olbrechts-Tyteca explain that these values are "traditional and accepted values . . . not new and revolutionary values" (1971, 51). When an arguer is able to collectively increase adherence to a value or set of values within the audience, they create communion with the audience and amongst its members. To promote communion, arguers can turn to "the whole range of means available to the rhetorician for purposes of amplification and enhancement" (1971, 51). Though there are a number of strategies for creating communion (Graff and Winn 2006, 54, 55, 57), Perelman and Olbrechts-Tyteca warn that there are no guarantees of success in persuasion, particularly because the value(s) the speaker/writer are asking the audience to commune with may be in conflict with other values that are important to the audience. They explain, "the argumentation in epideictic discourse sets out to increase the intensity of adherence to certain values which might not be contested when considered on their own but may nevertheless not prevail against other values that might come in conflict with them" (1971, 51). This suggests that even though the values invoked in epideictic argument may be widely shared or traditional, attempting to elevate them above other commonly held values can generate controversy.

If we accept the features identified by Perelman and Olbrechts-Tyteca as criteria for identifying epideictic, we should expect any number of these features to be present in real-world situations in which it occurs. There should be an audience whose resolution to act in the future is believed to need strengthening, and there should be attempts to strengthen that resolution by appealing to already commonly held beliefs or values. In considering the audience that McKellar targets in her books and the sociopolitical context from which the work emerged, we can identify these features. McKellar's primary audience is female students, in particular girls in middle school mathematics. That this demographic is the target for the book is made clear in her congressional testimony as well as in the book's title: *Math Doesn't Suck: How to Survive Middle School Math Without Losing Your Mind or Breaking a Nail.* That this audience's resolve requires strengthening is clear from the book's main title, *Math Doesn't Suck,* a counterargument to the audience's presumed perspective that it does. That this perspective represents a social problem or exigence that needs to be addressed has already been established in the previous section's discussion of the political effort dedicated to identifying the problem, its causes, and its potential solutions. The future action, of course, is to persuade young women and girls to pursue STEM degrees and join the workforce in STEM fields. Given that the text's audience, title, and context contain features we would expect from epideictic argument, it seems likely that further investigation would yield others, particularly attempts to create communion around and increase adherence to particular commonly shared values.

Math Doesn't Suck: Mathematical Epideictic for Young Women and Girls

With the epideictic genre and its features established as a proposed framework for analysis, we can now pursue answers to the questions "How does McKellar make mathematics appealing to her middle school audience?" and "What strategies does she use to persuade them to study and pursue careers in mathematics?" To address these queries, this section examines the layout, content, style, and argument in *Math Doesn't Suck*. This multifeature textual analysis suggests that McKellar appeals to a shared identity and argues from commonly held values.

Given that math is not a wildly popular topic for her audience, McKellar's first challenge is to attract and sustain their attention so that they will listen to and be predisposed toward her argument about the value of mathematics and mathematical careers. As Richard Graff and Wendy Winn point out, a number of strategies can be used to achieve these ends in epideictic argument, including referencing a common culture, tradition, or past, and coaxing the audience into actively participating with the speaker/writer in their exposition by taking them into their confidence, asking for their help, and identifying with them (2006, 55, 57). In *Math Doesn't Suck*, McKellar attempts to create a bond with her audience by strategically stylizing the layout, topics, and language of the book. These choices are designed to create a shared sense of feminine identity that McKellar hopes will attract young female readers to and sustain their interest in the mathematical methods and concepts she engages with in the text as well as encourage them to accept her arguments that they should positively valuate mathematics and take future action to pursue mathematically oriented educations and careers.

The most immediately obvious effort by McKellar to identify with her readers is in the design of the book's cover, which mimics the layout of popular teen and preteen magazines. The main cover image features McKellar striking a stylish pose in a fashionable outfit. Surrounding her image are pull quotes that would be instantly recognizable to young female readers familiar with popular teen magazines: "Do you still have a crush on him?" "Are you a math-o-phobe? Take this quiz!" "Horoscope inside!" In an interview with Ira Flatow, host of National Public Radio's (NPR) *Science Friday*, McKellar candidly admits using teen magazine conventions to coax her audience to open her book:

FLATOW   A horoscope section? Is this just an enticement to open the book
or—yeah.

MS. MCKELLAR   It's partially, absolutely. I modeled the book after a teen
magazine absolutely to make girls want to open the book. That's the first
step. Open it. (McKellar 2007a)

Once a reader has been lured inside the book, McKellar employs similar
strategies to keep her engaged in the explanations of mathematical concepts
that make up the core of the text.[20] To maintain the sense of a shared identity
with the reader as she moves through the book, McKellar engages with topics in
which she believes girls and young women would be interested, including shopping, fashion, pets, food, the arts (dancing, singing, and acting), romance, and
female friendship and rivalry. In chapter 5, for example, she draws on a number
of these themes in her illustration of how to multiply fractions: "Check this out:
Let's say you are in charge of hairdos for your school musical. Yeah, yeah. You
tried for the lead part but forgot the words in the middle of your song because
that guy you have a crush on was watching you. How embarrassing! . . . So,
there's a musical number with five girls and they all need to wear the same blue
ribbon in their hair. Each ribbon should be 2¼ feet long. How much total ribbon should you get at the fabric store?" (McKellar 2007b, 52).

In this word problem, McKellar draws on myriad themes to create a sense of
identification with her audience. The discussion of hairdos and musicals touches
on the topics of fashion and the arts, and the purchasing of the ribbon activates
associations with shopping. Also, a romantic crush on a boy sets the whole scenario in motion. By weaving relatable topics like these into mathematical discussions in the text, McKellar endeavors to hold the interest of her young female
readers by creating the vibe of a teen magazine.

In addition to strategically selecting topics that girls might be interested in,
McKellar also adopts an informal conversational style in her writing to encourage her readers to remain engaged in and feel comfortable with the material
she is covering. This style mimics the language choices and speech patterns
McKellar believes her female preteen/teen audience would use in their communications—a classical rhetorical strategy known as *ethopoeia*. To give her language the feel of being written by a preteen or teenage girl, McKellar adopts the
conventions of informal conversation, which include such features as the use of
contractions, slang phrasing, and digressions. All of these features are present in

the previously quoted word problem. The problem begins, for example, with the informal opening "Check this out." This phrase is in an informal slang register and is used instead of a more formal mode of address like "Think about this" or "Imagine this scenario." In fact, the imperative form of the phrase gives the text a conversational feel because it recognizes the need in conversation to gain an interlocutor's attention before entering into a verbal exchange with them. Typically, mathematical word problems do not attempt to connect with the reader in this way.[21] In fact, if we remove the imperative clause from McKellar's opening sentence, the initial sentence reads more like standard mathematical word problems that typically begin by establishing the context for the mathematical problem. By adding a conversational opening and using an informal register, McKellar gives the reader the sense that they are personally being invited by a female friend or confidant to consider a scenario in which they are cast as the mathematical protagonist.

McKellar also imitates informal conversation by including a digression in her word problem. We see an example of this in the second sentence of the problem, where McKellar interrupts the description of the problem with the phrase "Yeah, yeah" and then launches into a digression about why the reader is in charge of hairdos rather than starring in the play. In standard word problems, authors are unlikely to include digressions or to explain why the reader finds themselves in the hypothetical problem scenario. Such information is superfluous to solving the problem and may distract the reader from their mathematical task. For McKellar, however, this digression serves an important function. It gives her the opportunity to strengthen the identification between herself and her female readers by expressing sympathy that they didn't get the lead part in the play and by attributing their lack of success to a cause that she imagines her audience would relate to, a boy crush.

A detailed examination of the book's cover and a sample word problem illustrates the efforts McKellar makes in *Math Doesn't Suck* to attract women by creating a sense of identification between herself and her target audience of female middle school students. Encouraging young female readers to connect with the book is crucial for McKellar. If she can successfully capture their attention and predispose them to identify with the book's contents, she can encourage them to pay attention to the mathematical concepts in the text as well as prepare them to accept her main arguments: that mathematics is beneficial and that they should study and consider pursuing careers that involve it. Understanding how McKellar makes these arguments requires a close analysis of a variety of textual features, including the introduction; testimonials by students, TV celebrities, and professional women; and the "Reality Math" sections of the book.

The reason McKellar most frequently gives for why math is beneficial is that by studying it young women and girls can sharpen their intelligence. In the book's introduction, McKellar makes this point when she writes, "Working on math sharpens your brain, actually *making you smarter* in all areas. Intelligence is real, it's lasting, and no one can take it away from you ever" (McKellar 2007b, xvi). In the first sentence, McKellar clearly states a reason why math is beneficial. In the second, however, she pushes her conclusion further by arguing not only that mathematics improves intelligence, but also that intelligence itself is superior to other personal virtues. To make her case, she uses what Perelman (1982) refers to as arguments from dissociation (126–37) and double hierarchy (101–5). By claiming that intelligence is "real," McKellar dissociates it from other virtues that are only apparent and, therefore, not as highly valued as intelligence. In addition, McKellar also makes a double hierarchy argument for the superiority of intelligence. This argument relies on the assumption, presumably accepted by the audience, that phenomena that are permanent are superior to those that are fleeting. By connecting intelligence with permanence McKellar transfers the value of the permanent to intelligence, thereby reinforcing the superiority of intelligence over impermanent virtues.

McKellar uses similar arguments later in the introduction when she directly compares intelligence to the personal virtues of fame and beauty: "Take it from me, nothing can take the place of the confidence that comes from developing your intelligence—not beauty, or fame, or anything else '**superficial**.' . . . The good news is things that **really matter**, like our intelligence and personality— the things that feel good to be valued for—are things we have the ability to improve *ourselves*. While it's fun to focus on being fashionable and glamorous, it's also important to develop a smart and savvy side" (McKellar 2007b, xvii; emphasis mine). Here again we see McKellar valuating intelligence by appealing to dissociation. By naming beauty and fame "superficial" and intelligence "real," she asserts the superiority of intelligence over these other virtues. She also redeploys the double hierarchy argument; however, here she counts on the audience accepting as a warrant that virtues that can be cultivated or improved through personal effort are superior to those that cannot. Virtues like "intelligence and personality" which fall in the former category, she explains, are "things that really matter," while other virtues like beauty, fame, and glamour are "fun" but ultimately lesser virtues because they are not within our power to improve.

Though it may seem obviously suitable to elevate intelligence as a virtue when making a case for the value of mathematics, McKellar is in fact concerned that her audience may not accept this elevation because of what she believes is their disposition to be reluctant or ashamed to emphasize their intellectual abilities in peer social settings. In an interview with *ABC News* (2007) about the book, McKellar explains, "There is an epidemic right now of girls dumbing themselves down in middle school because they think it makes them attractive." To modify what she believes is her audience's disposition not to elevate their intelligence because they perceive males in their peer group will consequently devalue or undervalue their attractiveness, McKellar uses outside testimonials from their peers and role models in the "What Do You Have to Say?" sections of her book. In one of these sections, for example, McKellar quotes a seventeen-year-old student named Elyssa who openly challenges the practice of girls devaluing their intelligence: "I think a lot of girls dumb themselves down for boys. I don't see the point. I'm smart, and I also have a boyfriend. Besides, the guys you have to dumb yourself down for don't make good boyfriends anyway" (2007b, 50). With this brief quote, McKellar advances two arguments to reassure her female audience that they need fear no negative consequences to their attractiveness from elevating their intelligence. First, she argues that, in reality, there need not exist in the conceptual framework of suitable and attainable male love interests a competitive hierarchy in which intelligence and attractiveness compete. Elyssa's testimonial offers proof that young men without such competitive hierarchies exist. Second, McKellar reinforces her valuation of female intelligence by challenging the character of males who do do not value it. Their lack of recognition of intelligence as a value central to female virtuousness makes them fundamentally unsuitable romantic partners.

In other parts of the text, McKellar also relies on the testimony of young males, in particular TV stars, to debunk the stereotype that desirable young men consider intelligence and attractiveness as negatively reciprocal qualities. In a section titled "You've Seen Him on TV!," for example, McKellar quotes Devon Werkheiser, star of Nickelodeon's *Ned's Declassified School Survival Guide*, who testifies to the attractiveness of females who publicly express their intelligence. He explains, "I love smart girls. It always helps when a girl can hold up an intelligent conversation" (2007b, 8). With this quote, and others like it, McKellar reinforces the point that desirable young men are interested in and find intelligent females attractive. In aggregate, McKellar's use of testimonials from young women and men as well as her own explicit value arguments attempt to correct what she sees as a pernicious stereotype within and about her female audience. By arguing that feminine intelligence complements rather than competes with attractiveness in the value systems of suitable male romantic interests, she removes what she considers an obstacle to the appropriate valuation of intelligence. Once this appropriate valuation is restored, her audience's capacity to commit to the study of mathematics that demands public displays of intelligence will be improved.

Though encouraging her audience to reevaluate their beliefs about masculine stereotypes is an important dimension of McKellar's epideictic, she also makes appeals to the pragmatic benefits of mathematics. In particular, McKellar argues that mastering mathematics gives women the skills and confidence they need to tackle everyday challenges and improves their experiences in activities they enjoy. In the book's introduction, for example, she writes, "Math builds confidence, keeps you from getting ripped off, makes you better at adjusting cookie recipes, understanding sports scores, budgeting and planning parties and vacations, interpreting how good a sale really is, and spending your allowance" (2007b, xvi). Here, McKellar lays out for the reader a list of examples of how math can help them in their everyday lives that includes a broad range of activities, from protecting their economic interests to pursuing their personal passions for cooking, traveling, and sports. In the body of the book, McKellar reinforces this argument of utility in a variety of ways. One of the most common strategies she employs is to integrate into her word problems scenarios that her female readers might encounter in which understanding math would be useful. In chapter 7, "Is Your Sister Trying to Cheat You Out of Your Fair Share?," for example, McKellar illustrates the importance of being able to compare fractions to avoid being swindled out of a valued commodity, pizza. In her problem narrative, McKellar begins by setting up the problem scenario for the reader. Both she and her sister have ordered different pizzas that they intend to share. When the pizza arrives, the reader's sister realizes she has to leave, and they both realize that the pizzas have been divided differently: one into eight slices, the other into six. Once the problem scenario is laid out, McKellar asks her readers to consider how they might solve their problem with the help of mathematics: "Your sister offers you 2 of her ham and pineapple pieces ($2/6$ of a pizza), and you offer her 3 of your veggie pieces ($3/8$ of a pizza). She then complains, saying you should give her 4 of your pieces, because the pieces are so much smaller. Is she right—or could she be trying to cheat you out of your fair share?" (2007b, 75). With this final question hanging unanswered, McKellar dives into an explanation of how to compare fractions, the skill required to judge the fairness of the trade. By introducing real-world scenarios and asking the reader to figure them out on their own using mathematics, McKellar invites the reader to performatively experience the practicality of mathematics as a tool for dealing with everyday problems.

In addition to using unsolved word problems to persuade her readers of the practical utility of mathematics, McKellar also describes short real-world problems that she quickly solves with the help of math. These problems, which appear in "Reality Math" sections throughout the book, deal with a broad range of practical problems, from deciding whether the reader can afford to buy lunch for all of her friends to figuring out how many puppy videos she could fit on a DVD (2007b, 122, 136). In combination, McKellar's strategies of including repeated examples of the utility of mathematics and inviting her readers to apply mathematics to solve unfinished real-world problems reinforce the practical value of mathematics and add an additional force to her argument that mathematics is beneficial for them and should be pursued as a course of study.

Reshaping the image of mathematics so that middle school girls will find it valuable and relevant to their lives is an important aim for McKellar. The reasons she offers for why mathematics is beneficial serve as important stepping-stones toward her long-term goal of persuading her readers to devote their time and energy to studying mathematics and, eventually, pursuing careers in quantitatively oriented professions. That these are the critical concerns of the book is evidenced by McKellar's use of personal testimonials by women currently studying or working in mathematical fields. In these testimonials, she makes two persistent arguments to persuade her young female readers to take these future actions: (1) if you apply yourself to the study of mathematics, you can master it, and (2) studying mathematics can lead to an exciting, fulfilling, and/or lucrative career.

The argument that anyone who applies themselves to the study of mathematics can succeed is crucial for encouraging readers to continue their study of mathematics despite its challenges. To appreciate the significance of this argument, it is important to recall that *Math Doesn't Suck* is a reference book and, as a reference book, provides support and explanation of mathematical concepts for students currently struggling with mathematics. Given that the audience is perceived to struggle with the subject, encouraging them to persevere is crucial to keeping them on track to study mathematics in college or follow a career path toward a mathematical field. In the first testimonial, Jen Stern, a fourth-grade teacher from Los Angeles, makes an effort to connect with McKellar's audience of struggling readers by admitting that she too hated math and "was one of those kids who just didn't 'get it'" (McKellar 2007b, 25). In her testimonial, Stern details her tribulations and explains how she sought help from a female teacher who mentored her. Now she teaches math to fourth graders. At the end of her narrative, Stern leaves McKellar's readers with a simple message: "Don't give up, keep at it, and eventually, you will 'get it,' too!" Stern's message of encouragement is repeated in the second testimonial by petroleum analyst, actress, and web designer Stephanie Peterson, who, like Stern, struggled with math in school. Unlike Stern's experience, however, Peterson's illustrates how not devoting any effort to the study of math leads to bad consequences or lack of success. Peterson writes, "I didn't work very hard. . . . When I got my exams back [in middle and high school], I would cry because I had gotten a C or a D" (McKellar 2007b, 148). With the help of an inspiring female professor, however, Peterson turned things around. The professor noticed how hard Peterson was working in her class and how strong her grades were. She encouraged her to tutor, which led her to adding math as one of her majors. Her study of mathematics eventually resulted in a career in petroleum engineering. Like Stern's story, Peterson's narrative offers the reader struggling in math someone they can identify with. It also gives them hope that by applying themselves they too can begin to master mathematical ideas and operations.

Persuading readers to apply themselves in their mathematical studies is necessary to ensure that they put effort into mastering the mathematical concepts and methods they will need for success in school. However, the motivation to study hard also requires readers to have a sense of what the payoff will be for their hard work. The testimonials in McKellar's text also address this concern. In the second part of Stephanie Peterson's testimonial, for example, she describes her job as a petroleum engineer. In her description she recognizes the importance of mathematics to doing her job and highlights the qualities that make her career desirable. She begins with a statement of the utility of mathematics to her career: "Today, I use my math skills every day in my job as a petroleum analyst" (McKellar 2007b, 149). She then explains what makes her job so appealing: "What's so glamorous about all of this? Well, besides the fabulous job conditions (my own, beautiful office—yay!) and salary, I love what I do!" (2007b, 149). With these lines, Peterson creates an association for the reader between developing skills in mathematics and the benefits of a mathematically related career: good working conditions, high pay, and personal satisfaction.

Similar appeals both to the utility of mathematics and to the benefits of a career involving it also appear in the book's third testimonial, written by foreign exchange trader Tricia Hacioglu. Like Peterson, Hacioglu begins by describing her job and then explains what makes her work so compelling and how mathematics is important to doing it. She writes, "My job is exciting because I have to figure out how things that happen in the news are going to affect the currency rates. . . . I must constantly use logic and math. . . . At the end of each day, I calculate my daily profits and losses (hoping for lots of profits and no losses). . . . It's fast-paced and challenging, and I love it" (McKellar 2007b, 200). In describing her job, Hacioglu attempts to make her work appealing to the reader by suggesting that the tasks involved are exciting, that the job is potentially lucrative, and that she finds personal satisfaction in doing it. By supplying her young female audience with material and emotional incentives as well as female role models that they can identify with, McKellar attempts to persuade them to dedicate themselves to mastering mathematics and to seek out careers in fields like teaching, engineering, finance, and neuroscience.[22]

Analysis of the argumentative features of McKellar's text *Math Doesn't Suck* suggests that it has many of the hallmarks of the epideictic genre, including strategies to create identification, employ value arguments, and make gestures toward future action. In combination, these strategies interweave and support one another in McKellar's efforts to persuade middle school girls to engage with mathematics, understand its benefits, and explore educational opportunities and careers in it.

Audience Response to *Math Doesn't Suck*

Though a close reading of McKellar's book reveals myriad strategies for attracting middle school girls to mathematics, it does not provide insight into whether these strategies had the effects the author intended. Did it really convince middle school girls that math could be interesting, that they could succeed in it, or that they should pursue careers in it? Finding evidence of audience response is a notoriously tricky task; however, there are strategies available for gaining insight into some of these questions. To examine how McKellar's efforts were received by her intended audience and assessed by other cohorts of readers interested in promoting women's involvement in mathematics, this section analyzes reader reviews and critical online commentaries in traditional and social media on McKellar's book. These analyses suggest that though McKellar's epideictic did successfully connect with members of her target audience, her rhetorical strategies to identify with them invited criticism. By examining these criticisms as well as McKellar's defense of her book's strategies, this section illuminates the potential challenges that can arise from using female-specific appeals to attract young women and girls to mathematics and how McKellar defends her use of these tactics.

By standard metrics of popularity, McKellar's book was highly successful in attracting readers. After being released in July of 2008, the paperback edition of *Math Doesn't Suck* spent seven weeks on the *New York Times* Best Seller List, rising to as high as number three.[23] The book also received overwhelmingly positive customer reviews on Amazon.com and Barnesandnoble.com. On Amazon.com, for example, out of a total of 315 reviews, 75 percent rated the book five stars and 16 percent four stars, for a combined total of 91 percent of readers giving the book very positive or highly positive reviews. Reviews on BarnesandNobles.com, though not as overwhelmingly positive, also point to high audience satisfaction with the book. Out of 50 reviews on the site, 60 percent gave the book a five-star rating and 16 percent a four-star rating, for a combined total of 76 percent of reviewers giving ratings of the book that were very or highly positive.[24]

A qualitative assessment of these reviews suggests that McKellar's book was embraced by a broad spectrum of readers, not all of whom where middle school girls. In fact, many of the reviews were written by adults, both male and female, who found the book a great help in developing or refreshing their mathematical skills. There was also a critical mass of comments in the reviews by educators praising the book for the clarity and ingenuity of its explanations as well as its capacity to get students interested in mathematics. A comment left by "4th Grade Teacher" on Amazon.com captures the general spirit of posts left by educators: "I LOVE this book! Clearly, the book is aimed at young girls, not a 30 year old teacher, but the tips and tricks can be pulled out to work for all my students. For instance, when finding the Lowest Common Multiple, the textbook used prime factorization—something my students are struggling with. In this book, there is a 'birthday cake' method. My kids remember it and enjoy talking about cake in class" (4th Grade Teacher 2011).

Though general metrics suggest that McKellar's book was well received by a broad range of audiences, gaining insight into whether her target audience was attracted to the book requires a more detailed qualitative analysis. To get a sense of how middle school readers responded or were reported to have responded to the book, I analyzed a sample of the book's reviews on Amazon.com: 100 of the 286 four- and five-star reviews, and all 29 reviews with three or fewer stars. From this sample, I culled reviews in which the reviewer was either a female middle school student talking about her experience with the book or a parent or relative of a female middle school student reporting on their student's experience. Because the analysis of the book in the previous section focused on McKellar's efforts to identify with her readers and to persuade them to view mathematics positively, I singled out posts that commented specifically on these aspects of the readers' experiences. Other posts devoted to discussing the clarity or the comprehensiveness of the mathematical explanations in the book were disregarded because these features were more relevant to McKellar's efforts to make math intellectually comprehensible than to make it socially acceptable, though there is, no doubt, some overlap between these two outcomes. Using these criteria, I identified in my original sample of 129 texts twenty-six relevant positive comments on the book and two relevant negative comments.

A qualitative assessment of the positive reviews suggests that female middle school students liked, or were judged to like, McKellar's book and identified, or were judged to identify, with the choices of language and content she had strategically employed to attract them. In the sample set of twenty-six positive comments, half of them included statements about the book's likability. To make the case that middle school girls liked the book, reviewers oftentimes described the amount of time these girls spent with it and commented on their interest in exploring it on their own. Comments like the following were typical:

> The book arrived last week, and my daughter seems to always have her nose in it. (Kansas City Dale 2007)

> I gave it to her on a weekday in the morning. Six hours later she brought it up to me, already dog-eared and with sticky notes on many pages. (Novack 2012)

> I will have you know that since receiving it Tuesday, my daughter has not put the book down. She is VOLUNTARILY reading her new math book. (LadyJ "Janelle" 2014)

The reported attraction of these female readers to McKellar's math book suggests that her strategies for connecting with middle school girls had some positive influence on the target audience's reception to the book. However, these reviews do not provide specific evidence explicitly connecting the book's strategic choices of language and content to its positive reception. Other reviews, however, do make these connections. In one review, a twelve-year-old student recognizes and applauds McKellar's efforts to accommodate the book for girls when she writes, "Danica McKellar writes in a way that girls like me can relate to, so we understand how math is useful when we're grown-up. Although some people don't like it because she makes math sound girly, she's writing so that girls can both enjoy and learn from her book. I know a lot of girls who won't read good books because they aren't 'girl books.' *Math Doesn't Suck* is a math book and a girl book! I've already shown this to all of my friends, and my math teacher" ("I Love This Book!!!!" 2008).

In this passage, the reader makes a number of observations about *Math Doesn't Suck* that suggest that she recognizes and approves of McKellar's rhetorical strategies and embraces her goals for the book. In the first line, for example, she recognizes that McKellar's stylistic choices of language and content are conscious efforts to make the mathematical topics in the book relatable to girls. In the same line, she also identifies the author's pragmatic argument for why women should embrace mathematics. Interestingly, the line that follows reports a critique of the book that it "makes math sound girly." In this critique, we catch a first glimpse of controversy over the appropriateness of McKellar's efforts to feminize mathematics in response to the traditionally masculine stereotypes surrounding the subject. Despite encountering these critiques of the book's gendered approach, the reviewer closes her post by expressing her opinion on the importance and potential efficacy of McKellar's strategies to achieve her aims. She notes that there are "a lot of girls that won't read good books because they aren't 'girl books,'" implying that because McKellar's book is both a "good book" and a "girl book," it will attract girls and will have some benefit for them. By sharing the book with her friends, she reinforces her commitment to this belief that the strategies in the book will connect with girls and help them relate to mathematics.

Though reviews in which female students directly share their perspectives on the efficacy of McKellar's strategies are rare, there is a more substantial body of comments, twelve in total, by adult reviewers that provide further insight into the success of McKellar's efforts to create identification between herself and her target audience. In these comments, reviewers point to McKellar's strategic choices of language and content to explain why their female middle school student is attracted to the text:

> Both girls really enjoy this book. The author relates information in a way that the girls can understand and remember. It is almost *conversational in its explanations.* (clcPsalm19 2007; emphasis mine)

> Our granddaughter was looking forward to receiving the book if for no other reason than it has a *cool title.* (Old Guy 2007; emphasis mine)

> The added *stories/notes for girls* makes my daughter like reading her math book. (BellaMaurita 2008; emphasis mine)

> *She loves teen magazines, gossip, shopping, girly stuff* and the likes and *McKellar's books really appeal to her*—a MATH book that she can pick and read for fun and learn something in the process. (MH "maryham" 2011; emphasis mine)

These responses highlight a number of features described earlier in the rhetorical analysis as strategies used by McKellar to create a sense of identification with her readers. The first quote, for example, highlights the conversational style in the text, while the second homes in on its use of informal slang. The second and third quotes include statements about how choices in subject matter (gossip, shopping, and stories/notes for girls) and genre (teen magazine) also attracted the attention of young female readers.

Though the responses that highlighted and evaluated the rhetorical strategies in McKellar's book were overwhelmingly positive, there were a few negative reviews that criticized the strategic choices she used to connect with her female audience. Authors of these critiques found McKellar's strategies distracting and insulting to female readers. One fourteen-year-old reviewer wrote, for example, "All I got out of this book was a bunch of lipstick, lip gloss, and poodle-based ways of figuring out math problems, which did not help me at all, as I am serious with my work and my grades" (Palimino 2013). The father of another student of a similar age reported, "The rising seventh grader says they [McKellar's books] were 'somewhat condescending, [and] tried too hard to be appealing' by referencing pop icons who she cares nothing about and 'who will be out of vogue in a couple years'. . . . It seemed to us the whole pop culture thing appeared very forced, awkward, and ultimately distracting" (Espresso Pete 2010).

In these reviews, the authors perceived McKellar's rhetorical strategies as distractions from the more serious, unrelated mathematical content of the text. For the first, presumably female, reviewer, McKellar's efforts to undo the antithesis between attractiveness and mathematical intelligence by including "a bunch of lipstick, [and] lip gloss" negatively influenced her reception of the text. In the second review, the female middle school student is reported to have found McKellar's rhetorical efforts to connect with her both inappropriate and unsubtle, impeding her efforts to engage with the mathematical content of the book and insulting her intelligence by trying "too hard to be appealing."

Defending a Feminine Approach to Mathematics

Aside from a few negative reviews, the majority of middle school girls in online comments responded or were reported to have responded favorably to McKellar's book. These responses suggest that McKellar's rhetorical efforts at generating interest in mathematics and creating a sense of identification between herself and her intended audience were reasonably successful with this cohort of readers. As the negative comments suggest, however, there was less than a unanimous opinion among her target audience that McKellar's strategically feminine approach to mathematics was appropriate or conducive to creating a sense of identification with her female audience. These critics were not alone in their objections. Commentators in mainstream and online media were also critical of McKellar's mathematical epideictic. Exploring their objections more clearly illuminates different perspectives about the appropriateness of appealing to feminine values and virtues to attract girls and young women to mathematics.

An online search of reviews of the book and interviews conducted with McKellar about it generated eight sources in which McKellar addressed questions about or criticisms of her strategies.[25] In these sources, a number of common themes emerged. Among these was that McKellar's choice of strategies for identifying with her female audience advanced or embraced an image of femininity that the respondent felt was regressive. In an interview with Aaron Rowe of *Wired*, for example, Rowe asks McKellar, "There are a lot of references to baking cookies, expensive clothing, cosmetics and accessories. This could be viewed as fun—or reinforcing gender roles?" (Rowe 2007). Here Rowe suggests that McKellar's rhetorical strategies perpetuate traditional masculine stereotypes of feminine behavior that he finds regressive. A more forceful but similarly directed critique against the book is leveled by Carol Lloyd, a contributor to Salon.com's feminist blog Broadsheet, who writes, "Can you really motivate girls to use their minds via their Miu Miu ostrich satchel fantasies? Should we even try to? Don't get me wrong, I hope the book is a giant hit, spawning a rash of mathematics-and-manicure slumber parties the likes of which the country has never known. . . . But I have my doubts. When it comes to educating girls in something as unsexy as math, trotting out sexist formulas may seem practical, but it's impossible to calculate what else we're subtracting" (Lloyd 2007).

Like Rowe, Lloyd takes McKellar to task for "trotting out sexist formulas" on the grounds that it may be subtracting, presumably, from current efforts in feminism to move women away from identifying with materialism, shopping, and other traditionally feminine stereotypes. Although she recognizes that McKellar is appealing to these stereotypes in the hopes of attracting girls who have had a hard time developing an interest in math, she also argues that "the book sounds like it walks a fine line between endorsing inane stereotypes and using them to lure girls into learning their numbers" (Lloyd 2007).

In her interactions with the media, McKellar is cognizant of these critiques and defends herself from charges that she is promoting regressive feminine stereotypes by arguing that she is trying to reach and motivate a particular demographic of girls who already embrace these perspectives on femininity rather than to perpetuate unhelpful stereotypes. McKellar adopts this stance in her response to Rowe's question which she answers in the following fashion: "What do you think? If I'm teaching girls that do love to make cookies and do love fashion—you think that's me saying, come on girls you belong in the kitchen, you belong shopping? Or do you think it's me showing them how math is part of all their life, even the part they thought it had nothing to do with?" (Rowe 2007). With these rhetorical questions, McKellar challenges the assumptions made by Rowe in his question about her motivation for choosing the subjects in the book. She asks him whether he believes she is trying to persuade girls to adopt a regressive view of femininity or whether he believes her motivation is to use these subjects to help girls identify with mathematics. Further, when Rowe (2007) asks McKellar in an earlier portion of his interview, "Did you talk to any girls who have read the book? What kind of contact have you had with junior high school girls?," McKellar describes her efforts to get to know her audience, suggesting that her choices in the book were audience-informed rather than dictated by her own perspectives on female identity.

In addition to defending her choices in the book by arguing that she was meeting girls where they are, McKellar also maintains that women should be allowed to define themselves as they please and not be restricted by any particular principle of femininity. She raises this defense in an interview with Tara Smith, founder of Iowa Citizens for Science. In her interview with McKellar on her blog *Aetiology*, Smith offers a critique of the book similar to Rowe's and Lloyd's when she asks McKellar, "In the book's introduction (and in many places throughout), you emphasize brains over beauty. . . . Yet many of the examples used in the book reinforce the stereotypes—loving diamonds, shoes, make-up, and shopping . . . are you sending mixed signals here?" (Smith 2007). In her question, Smith acknowledges and applauds what she considers McKellar's progressive message about femininity that emphasizes the importance of brains over beauty. However, this message, in Smith's estimation, is contradicted by McKellar's focus in the text on fashion and materialism that she believes return to a regressive perspective on femininity by reversing the hierarchy of these values. McKellar responds that Smith's progressive/regressive distinction between intelligence and beauty is a false dichotomy. Both, she argues, are important aspects of the character of the modern woman:

> The fact that it's not a mixed signal is exactly the central message of my book: Girls can enjoy being "girly" and "fabulous" alongside developing their brain. . . . I don't want girls to feel like they are being boxed into stereotypes—nerdy, superficial, or otherwise. Let them define the young woman they want to be, and if being attractive and fashion savvy is one of their goals . . . then the most empowering thing I can do for girls . . . [is] to show them how being smart is an essential element to the young woman they are training to be. (Smith 2007).

Here McKellar defends here choices in the book by suggesting that Smith is herself perpetuating regressive, essentialist stereotypes when she claims that intelligence and beauty cannot be coelevated as female values. By allowing that both can be components of a complex female character, McKellar supports a more comprehensive notion of femininity embracing women of all kinds as they pursue their intellectual development in mathematics.

Interestingly, this same argument appears in defenses of McKellar's book in cases where the author herself is not interviewed, suggesting that it represents a generally held perspective on the role of fashion, beauty, and materiality within the modern feminine ideal and their relationship to the intellect or intellectual development. In online comments on Lloyd's review of *Math Doesn't Suck*, for example, several respondents chastise the feminist blogger for applying the stereotyped division between intelligence and beauty to criticize McKellar's efforts. One respondent identifying themselves as 5656565656, for example, writes:

> Yeah, I do really wonder about the contempt directed at "girliness." Is it that Broadsheet readers think that wearing makeup makes you stupid? Or wearing nice clothes? Or jewelry? The thing is, if we, as feminists, continue to believe that being girly is the same as being stupid, it will become a self-fulfilling prophecy. A girl who is into clothing and makeup (and many girls are) will, after absorbing this stereotype, start believing that she is stupid and that she can't do math—and perpetuate the stereotype. So instead of criticizing Danica McKellar, we should applaud her attempt. Or is it that she's wearing too much makeup to be taken seriously? (5656565656 2007)

In these lines, the respondent takes Lloyd to task on the blogger's division between intelligence and behaviors associated with feminine attractiveness by challenging the logic of their antithetical relationship to one another. If being attractive and intelligent are opposites, then does being attractive make you stupid? Is attractiveness or unattractiveness a reliable sign of a woman's intelligence? These rhetorical questions (to which the respondent expects the audience to answer "no") challenge the strength of the cause/consequence relationship between feminine attractiveness and intelligence that the respondent believes Lloyd is suggesting with her comments. In the closing rhetorical question, the respondent even challenges Lloyd to apply her standard to McKellar, who clearly is both beautiful and intellectually gifted in mathematics. By challenging the assumed oppositional character of these qualities, the author also highlights the negative consequences of believing these qualities are mutually exclusive. Not seeing them as compatible, the respondent argues, means women who are interested in their attractiveness will be unfairly stereotyped as dumb and may lose confidence in their intelligence and give up on the study of math altogether.

For many respondents, young women and girls giving up on math or feeling alienated from it was a far worse consequence than their being exposed to story problems about shopping or boy crushes. Though reviewers themselves did not always identify with McKellar's female-centered strategies for attracting her audience to math, they believed that the audience she was targeting, middle school girls, likely would identify with them and would ultimately benefit from her efforts to attract them to mathematics. Responses like the following illustrate this perspective:

> I may not like the fact that little girls tend to be into shopping, shoes, and handbags, but it is reality, and I'd rather it not doom them to a life of nothing BUT shopping, shoes, and handbags. A love of math will allow them to move beyond this world, whether they know it or not. (jbldmm 2007)

> While I don't feel that books like these are working to deconstruct the oppositional binary [between intelligence and beauty] . . . I feel like we also can't discount targeting a pink demographic. . . . Getting more girls and women into STEM fields is an acute problem. . . . If we don't find a way to target teen girls with math skills now, that is another generation of girls told that they can't. (Feminist Salarian 2012)

In each of these comments, the respondents, whether they find McKellar's approach to her audience optimal or not, agree that the demographic that she is trying to reach exists and is sufficiently large or important to warrant being addressed. Further, all of them recognize McKellar's goal as a worthy one and consider her ends to justify the means she is employing to attain them.

Psychological Research and Theories of Audience

An examination of the critical comments and questions about McKellar's mathematical epideictic and responses to these comments suggests divided evaluations and interpretations of the rhetorical choices she makes in the text. Respondents who evaluated the text positively interpreted her feminization of mathematics as an effort to counter stereotypical perceptions that girls interested in typically feminine pursuits could not also be interested in or good at mathematics. Reviewers who characterized it negatively, however, believed that her attempt to attract young women to mathematics by appealing to their materialism, obsession with pop culture, or vanity would affirm regressive feminine stereotypes. Adjudging the appropriateness of these evaluations of *Math Doesn't Suck* is a complex task beyond the scope of evidence and methods of analysis in this chapter. However, a brief examination of the sociological and psychological literature that explores the relationship between gender stereotyping and female mathematical performance suggests that the perspectives of both McKellar's supporters and critics find some confirmation in empirical research, though not in all cases for the reasons on which they base their evaluations.

Work in social psychology suggests that the problem of masculine stereotyping in mathematics that McKellar and her supporters argue is the raison d'être for her rhetorical choices is a very real concern. In a landmark study, *Talking About Leaving: Why Undergraduates Leave the Sciences*, sociologist Elaine Seymour and women's studies professor Nancy Hewitt report that feminine behaviors that contradict the stereotypes of science, math, and engineering (s.m.e.) as a masculine pursuit, like dressing attractively or wearing makeup, evoked negative stereotyping in male students and raised fears in female students about being judged negatively. Summarizing focus group discussions with male students who had dropped out of s.m.e. majors, the authors recount: "They [male undergraduates] tended to view any woman's interest and ability in science and mathematics as 'unnatural,' they were apt to portray women who chose s.m.e. majors . . . as inherently ugly, as having been too busy with academic work to learn the arts of attractive self-presentation; as having lost their attractiveness after they entered the sciences" (Seymour and Hewitt 1997, 248). Subsequent research by Pronin, Steele, and Ross, who interviewed female students in quantitatively focused courses of college study, supports these findings, confirming that female students perceived masculine stereotypes about the relationship between feminine attractiveness and stem study, rating "wearing of makeup" and "fashion consciousness" as behaviors that they believed could provoke their male classmates to negatively stereotype them (Pronin, Steele, and Ross 2004). The evidence of male stereotype threat provided by these studies supports the interpretation/evaluation of McKellar's proponents that her feminization of mathematics was strategically designed to combat prevailing masculine stereotypes rather than, as her detractors maintain, to perpetuate them. In fact, McKellar herself signals this strategic intention in her response to Tara Smith's questioning of her feminization of math when she responds, "Girls can enjoy being 'girly' and 'fabulous' alongside developing their brain. . . . I don't want girls to feel like they are being boxed into stereotypes" (Smith 2007).

In addition to providing support for McKellar's and her proponents' interpretation of her rationale for feminizing mathematics, the psychological research also offers perspective on the suitability of her female-centered rhetorical strategy to address middle school–aged girls' lack of interest in mathematics, fear of being negatively stereotyped, and reluctance to pursue further study or careers in mathematics. While the reviews of the book posted on the Amazon and Barnes and Noble websites suggest that McKellar had some success in getting female students interested in mathematics, these reviews offer no evidence of the book's capacity to address the other problems. Research in social psychology, however, raises questions about whether McKellar's strategy of using glamorous female role models to inspire young girls to study math would achieve the positive outcomes she intends. In "My Fair Physicist? Feminine Math and Science Role Models Demotivate Young Girls," Diana Betz and Denise Sekaquaptewa (2012) explore whether feminine STEM role models motivate or demotivate middle school girls from perusing STEM careers. To test the hypothesis that feminine role models would demotivate girls, they asked 193 sixth- and seventh-graders to read magazine-type interviews about female university students, some of whom were dressed in feminine styles while others wore gender-neutral fashions. After reading these interviews, the participants were asked to rate the role models for positive attributes and the degree of identification they had with them, as well as the participants' own future plans to study mathematics.

While Betz and Sekaquaptewa found no statistically significant difference between students' capacity to identify with or feel positively about feminine STEM role models, they discovered that this type of role model had a demotivating influence on STEM-disidentified students' future plans to study math. In a second study, they isolated what they believed were the psychological factors behind this demotivation, concluding that "role models who represent an unattainable standard make audiences feel threatened rather than inspired" (2012, 743). In Betz and Sekaquaptewa's estimation, feminine STEM role models constituted an unattainable standard for middle school girls who do not already identify with STEM because it required them to accept an improbable future in which they both succeeded at mathematics and embodied the virtue of feminine attractiveness. They explain that, "at an age when stereotypes about gender and scientists are rather rigid, being a feminine woman in STEM may seem particularly unlikely" (2012, 739). Gender-neutral female role models, however, were more likely to be judged by STEM-disidentified students to represent feminine standards they could attain.

Betz and Sekaquaptewa's study raises some intriguing questions about whether some of McKellar's rhetorical choices are optimal for achieving her persuasive goals. Her epideictic strategy, for example, of profiling feminine STEM role models like Stephanie Peterson, the petroleum engineer, and Tricia Hacioglu, the foreign exchange trader, may demotivate rather than inspire her intended audience to pursue mathematical courses of study because of their impressive feminine beauty and math smarts. Perhaps more seriously, McKellar herself, an attractive TV star with a degree in mathematics, may present her readers with an imposing standard, undermining her effectiveness to be a role model for her primary audience. In neither case, however, does the study suggest that girls reject these role models for being regressive stereotypes for femininity. On the contrary, the studies suggest that these role models transcend middle school girls' typical stereotyping of femininity to such a degree that the juxtaposition of femininity and mathematics seems so unrealistic as to be unattainable.

Conclusion

By examining the text, contexts, and audience responses to *Math Doesn't Suck*, this chapter has endeavored to contribute to rhetorical and feminist studies of mathematics by examining the female-centered strategies used by Danica McKellar to attract middle school girls to mathematics, and the responses these strategies received from her audiences. A close textual analysis employing an epideictic framework revealed that McKellar employs informal registers, value arguments, and female role models to attract the attention of, create a sense of identification with, and inspire future action in her readers. Consequent assessment of reader responses showed that while McKellar's strategies succeeded in attracting a substantial cohort of her target audience to the text, her efforts to feminize mathematical study garnered conflicting evaluations and interpretations from her reviewers. While critics characterized her rhetorical tactics as pandering to regressive stereotypes, her defenders viewed them as efforts to combat them. Further contextualization of her work using recent psychological research suggested that the problems of gender stereotyping that McKellar claims to be addressing exist and constitute a serious problem for female students. However, the research also raised questions about the potential efficacy of using female role models to inspire middle school girls not already identifying with mathematics to pursue further study or career paths in it.

The work in psychology, particularly the questions it raises about the potential efficacy of rhetorical interventions to address gender stereotypes within certain audiences, invites further research and transdisciplinary collaboration between psychologists, sociologists, and rhetorical scholars. Betz and Sekaquaptewa's research, for example, suggests that a more nuanced examination of the factors that might influence the rhetorical efficacy of role models needs to be undertaken. They show that in cases where role models were strategically deployed to counter negative stereotypes, their capacity to influence audience perspectives and imagined future behavior was limited by the intensity to which that audience adhered to the stereotype the role models were trying to disconfirm (2012, 743). Because role models have been widely embraced as a strategy for attracting female students to STEM, further engagement with this literature might provide additional insights into how audience predispositions influence their reception of role models and how the features of role models might activate different responses in audiences. We might inquire, for example, whether exposing male audiences to feminine STEM role models might reduce negative stereotyping, and, if so, whether this sort of exposure could lower the stereotype threat experienced by female students in academic settings.

Whereas rhetorical scholars might gain important insights from educators and psychologists into the psychological characteristics of different audiences, they might supply researchers in these areas with a more nuanced understanding of the different modes of representation and strategies available for countering stereotypes. Betz and Sekaquaptewa's analysis, for example, focuses exclusively on assessing the impact of visual cuing (role models dressed in feminine styles) as a strategy for countering the stereotype that STEM is an unfeminine field of study. As the analysis in this chapter of McKellar's rhetorical effort in *Math Doesn't Suck* reveals, however, a variety of other strategies for counterstereotyping are also available, including testimonials from middle school girls and desirable male TV stars. Additionally, McKellar infuses the study of mathematics with female middle school culture and language, performing in the text a conceptual blend that invites her readers to develop new mental models that contradict current stereotypes of mathematics as unfeminine. In future research, scholars in rhetoric and psychology might collaborate to catalogue the diversity of counterstereotyping strategies available and test their efficacy in weakening female students' adherence to negative gender stereotypes.

Currently, there are many more questions than answers about the capacity of gendered rhetorical strategies to address the social-psychological challenges that girls and women face in their mathematical education and on their path to careers in quantitative fields. This rhetorical analysis of McKellar's text and the reader responses to it has endeavored to identify strategies intended to attract female readers and to assess their influence on primary and tertiary audiences of the text. Though in rhetoric, as in most things in life, there is no guarantee of success, efforts to more thoroughly understand the available means of persuasion and the characteristics of the audiences and contexts in which they are efficacious can help set us on the path to making mathematics a safer and more desirable intellectual space for women and girls to inhabit.

Notes

1. There have been discussions in the psychological literature of interventions meant to address the problem of stereotype threat for females in mathematics. These interventions, however, are all nonliterary, involving, for example, the introduction of female test proctors (Marx and Roman 2002) or in-group expert role models (Stout et al. 2011).

2. *Hearing on Women in Science and Technology Equal Opportunity Act, Before the Senate Subcommittee on Health and Scientific Research*, 96th Cong. 1 (1980) (statement of Howard Metzenbaum, Senator from Ohio).

3. Ibid., 12.

4. *Hearing on Women, Minorities, and the Disabled in Science and Technology, Before the House Subcommittee on Science, Research, and Technology*, 100th Cong. 1 (1988) (statement of Doug Walgren, Chairman of the Subcommittee on Science, Research, and Technology).

5. *Hearing on Women in Science and Technology Equal Opportunity Act, Before the Senate Subcommittee on Health and Scientific Research*, 96th Cong. 4 (1980).

6. Ibid., 6.

7. Ibid.

8. Ibid., 6–7.

9. Ibid., 8.

10. In 1980 approximately 42.3 percent of bachelor's degrees were earned by women, which increased to 45.8 percent in 1996 (NSF 2000, 119). Women accounted for 36.1 percent of all master's degrees earned in 1980 and 40.2 percent in 1996 (NSF 2000, 170).

11. *Hearing a Review of the Morella Commission Report: Recommendations to Attract More Women and Minorities into Science, Engineering, and Technology, Before the House Subcommittee on Technology*, 106th Cong. 1 (2000).

12. Ibid., 7.

13. *Hearing on Women, Minorities, and the Disabled in Science and Technology, Before the House Subcommittee on Science, Research, and Technology*, 100th Cong. 41 (1988) (statement of Danica McKellar, witness).

14. Ibid.

15. Ibid., 46.

16. Ibid., 47.

17. Ibid., 45.

18. Ibid., 49.

19. Articles by Majdik, Platt, and Meister (2011) and Little and Branker (2012) are exceptions here, though they explore audience in only a general sense, without specific attention to gender.

20. In this analysis, I am attributing the majority of strategic content choices to McKellar rather than to an editor or publisher. I make this assumption on the basis of McKellar's own admission that she was primarily responsible for the text. In an interview with *Wired Magazine*, she states, "There was almost no help. The publisher I used is not a textbook publisher— it's a young adult's publisher. They do novels. I got comments from them sometimes, but not much" (Rowe 2007).

21. In her book on mathematical word problems as a genre *A Man Left Albuquerque Heading East*, Susan Gerofsky identifies three standard generic features of word problems: (1) a setup establishing the characters and location of the story, (2) the information needed to solve the problem, and (3) a question (Gerofsky 2005, 27).

22. The last testimonial, not analyzed here, was from an undergraduate who was majoring in neuroscience at the University of Southern California (McKellar 2007, 255–56).

23. The book was on the list from the week of August 24 to the week of October 5, 2008, and was number three on the week of September 28, 2008 ("Math Doesn't Suck" 2008).

24. The analysis of reviews was done in June of 2015.

25. In order to locate sources, I did a Google search using the phrases "Danica McKellar" and "Math Doesn't Suck." I also used McKellar's publicity website for the book (http://www .mathdoesntsuck.com/reviews/#top) to identify additional commentaries.

References

BellaMaurita. 2008. Review of *Math Doesn't Suck*, by Danica McKellar. Amazon, August 24. https://www.amazon.com/gp/customer-reviews/R1UE4T3ODNHKEJ.

Betz, Diana E., and Denis Sekaquaptewa. 2012. "My Fair Physicists? Feminine Math and Science Role Models Demotivate Young Girls." *Social Psychological and Personality Science* 3 (6): 738–46. https://doi.org/10.1177/1948550612440735.

Chang, Kenneth. 2005. "Between Series, an Actress Became a Superstar (in Math)." *New York Times*, July 19. https://www.nytimes.com/2005/07/19/science/between-series -an-actress-became-a-superstar-in-math.html.

Clcpsalm19. 2007. Review of *Math Doesn't Suck*, by Danica McKellar. Amazon, September 15. https://www.amazon.com/gp/customer-reviews/R296NB2I0Y1DKT.

Damarin, Susan. 2008. "Toward Thinking Feminism and Mathematics Together." *Journal of Women in Culture and Society* 34 (1): 101–23. http://www.jstor.org/stable/10.1086 /588470.

Espresso Pete. 2010. Review of *Math Doesn't Suck*, by Danica McKellar. Amazon, August 2. https://www.amazon.com/gp/customer-reviews/R1N1ZBIX4FV4QX.

Fahnestock, Jeanne. 1998. "Accommodating Science: The Rhetorical Life of Scientific Facts." *Written Communication* 15 (3): 330–50. https://doi.org/10.1177/0741088386003003001.

Feminist Salarian. 2012. Comment on "Is Math Book Bad for Girls?" Patheos, August 18. http://www.patheos.com/blogs/friendlyatheist/2012/08/18/is-this-math-book -bad-for-girls.

565656565656. 2007. "What Is Wrong with Applying Math to Real Life?" (Comment). Salon, August 6. https://www.salon.com/2007/08/06/math.

4th Grade Teacher. 2011. Review of *Math Doesn't Suck*, by Danica McKellar. Amazon, October 27. https://www.amazon.com/gp/customer-reviews/R1FXNSWDRGYP7S.

Gerofsky, Susan. 2005. *A Man Left Albuquerque Heading East: Word Problems as Genre in Mathematics Education*. New York: Peter Lang.

Graff, Richard, and Wendy Winn. 2006. "Presencing 'Communion' in Chaïm Perelman's *New Rhetoric*." *Philosophy and Rhetoric* 39 (1): 45–71. https://doi.org/10.1353/par.2006.0006.

Harding, Sandra. 1986. *The Science Question in Feminism*. Ithaca: Cornell University Press.

Hayden, Wendy. 2013. *Evolutionary Rhetoric: Sex, Science, and Free Love in Nineteenth-Century Feminism*. Carbondale: Southern Illinois University Press.

Henrion, Claudia. 1997. *Women in Mathematics: The Additional Difference*. Bloomington: University of Indiana Press.

"I Love This Book!!!!" 2008. Review of *Math Doesn't Suck*, by Danica McKellar. Amazon, January 3. https://www.amazon.com/Math-Doesnt-Suck-Survive-Breaking/product-reviews/0452289491 (review no longer active).

Jack, Jordynn. 2009. *Science on the Home Front: American Women Scientists in World War II*. Carbondale: Southern Illinois University Press.

Jbldmm. 2007. "As a Woman Who Loves Math …" (Comment). Salon, August 6. https://www.salon.com/2007/08/06/math.

Kansas City Dale. 2007. Review of *Math Doesn't Suck*, by Danica McKellar. Amazon, August 10. https://www.amazon.com/gp/customer-reviews/R1CELCNA1RW0LO.

Keller, Evelyn Fox. 1995. *Refiguring Life: Metaphors in Twentieth-Century Biology*. New York: Columbia University Press.

LadyJ "Janelle." 2014. Review of *Math Doesn't Suck*, by Danica McKellar. Amazon, July 3. https://www.amazon.com/gp/customer-reviews/R6SY9HXBKB48Z.

Little, Joseph, and Maritza Branker. 2012. "Analogy in William Rowan Hamilton's New Algebra." *Technical Communication Quarterly* 21 (4): 277–89. https://doi.org/10.1080/1057 2252.2012.673955.

Lloyd, Carol. 2007. "Math Doesn't Suck, It Buys You Gucci." Salon, August 6. https://www.salon.com/2007/08/06/math.

Majdik, Zoltan, Carrie Anne Platt, and Mark Meister. 2011. "Calculating the Weather: Deductive Reasoning and Disciplinary *Telos* in Cleveland Abe's Rhetorical Transformation of Meteorology." *Quarterly Journal of Speech* 97 (1): 74–99. https://doi.org/10.1080/00335630.2010.539622.

Marx, David M., and Jasmin S. Roman. 2002. "Female Role Models: Protecting Women's Math Test Performance." *Personality and Social Psychology Bulletin* 28 (9): 1183–93. https://doi.org/10.1177/01461672022812004.

"Math Doesn't Suck." 2008. *New York Times Best Sellers*, August 21 to October 5. https://www.nytimes.com/books/best-sellers/2008/10/05/paperback-books/?smid=em-share.

McKellar, Danica. 2007a. Interview by Ira Flatow. *Talk of the Nation: Science Friday*, National Public Radio, September 21. https://www.npr.org/templates/transcript/transcript.php?storyId=14594340.

———. 2007b. *Math Doesn't Suck*. New York: Plume.

MH "maryham." 2011. Review of *Math Doesn't Suck*, by Danica McKellar. Amazon, October 21. https://www.amazon.com/gp/customer-reviews/R3EV8RDCOC7IMX.

Murray, Margaret. 2000. *Women Becoming Mathematicians: Creating a Professional Identity in Post–World War II America*. Cambridge, MA: MIT Press.

Novack, S. C. 2012. Review of *Math Doesn't Suck*, by Danica McKellar. Amazon, August 15. https://www.amazon.com/gp/customer-reviews/R21UZA5OO0OJP1.

NSF (National Science Foundation). 2000. *Women, Minorities, and Persons with Disabilities in Science and Engineering: 2000.* Arlington, VA: NSF. http://archive.cra.org/Activities /workshops/broadening.participation/nsf/wmd00.pdf.

Old Guy. 2007. Review of *Math Doesn't Suck*, by Danica McKellar. Amazon, September 4. https://www.amazon.com/gp/customer-reviews/RXSH1RHJLOO8G.

Palimino. 2013. Review of *Math Doesn't Suck*, by Danica McKellar. Amazon, February 20. https://www.amazon.com/gp/customer-reviews/R1OF63BN8TK0BX.

Perelman, Chaïm. 1982. *The Realm of Rhetoric.* Notre Dame: University of Notre Dame Press.

Perelman, Chaïm, and Lucie Olbrechts-Tyteca. 1971. *The New Rhetoric.* Notre Dame: University of Notre Dame Press.

"Person of the Week: Danica McKellar." 2007. ABC News, August 10. https://abcnews.go .com/WN/PersonOfWeek/story?id=3467211.

Pronin, Emily, Claude M. Steele, and Lee Ross. 2004. "Identity Bifurcation in Response to Stereotype Threat: Women in Mathematics." *Journal of Experimental Social Psychology* 40:152–68. http://doi.org/10.1016/S0022-1031(03)00088-X.

Rowe, Aaron. 2007. "Math Book Helps Girls Embrace Their Inner Mathematician." Wired, August 2. https://www.wired.com/2007/08/math-book-helps-girls-embrace-their -inner-mathematician.

Schiebinger, Londa. 1989. *The Mind Has No Sex: Women and the Origins of Modern Science.* Cambridge, MA: Harvard University Press.

———. 2008. *Gendered Innovations in Science and Engineering.* Stanford: Stanford University Press.

Seymour, Elaine. 1995. "The Loss of Women from Science, Mathematics, and Engineering Undergraduate Majors: An Explanatory Account." *Science Education* 79 (4): 437–73. https://doi.org/10.1002/sce.3730790406.

Seymour, Elaine, and Nancy M. Hewitt. 1997. *Talking About Leaving: Why Undergraduates Leave the Sciences.* Boulder: Westview Press.

Smith, Tara. 2007. "Interview with Math Whiz, Author, and Actress Danica McKellar." Aetiology (blog), July 25. http://aetiologyblog.com/2007/07/25/interview-with-math -whiz-autho.

Spencer, Steven, Claude Steele, and Diane Quinn. 1999. "Stereotype Threat and Women's Math Performance." *Journal of Educational Social Psychology* 35 (1): 4–28. https://doi .org/10.1006/jesp.1998.1373.

Stout, J. G., Nilanjana Dasgupta, Matthew Hunsinger, and Melissa A. McManus. 2011. "STEMing the Tide: Using Ingroup Experts to Inoculate Women's Self-Concept in Science, Technology, Engineering, and Mathematics (STEM)." *Journal of Personality and Social Psychology* 100 (2): 255–70. https://doi.org/10.1037/a0021385.

9

Turning Principles of Action into Practice | Examining the National Council of Teachers of Mathematics' Reform Rhetoric

Michael Dreher

Within the field of public education, questions often arise about what books should be read in school, which parts of our country's history should be taught, or what kinds of science are most valuable. In mathematics, as in these other subjects, public deliberations are no different. Parents, scholars, and other interested parties have strong opinions about what students should learn and how they should learn it. The National Council of Teachers of Mathematics (NCTM) is often at the forefront of educational reform debate as "the organization that launched the education standards movement" and the self-proclaimed "public voice of mathematics education" (NCTM 2006, iv). Given the NCTM's history of being involved in educational reform and the creation of standards like the Common Core, it provides a useful model for considering how a professional organization influences public discourse about mathematics and its teaching. Unlike the subjects of science, history, and English, where there has been fragmentation amongst professional educational organizations, the NCTM has remained the primary advocacy organization for curricular and pedagogical reform in mathematics. As a result, the NCTM has a national footprint and considerable influence over public disputes involving mathematics curriculum and pedagogy. In this chapter, I examine one of the more recent controversies over the Common Core standards.

This chapter describes how the National Council of Teachers of Mathematics operates within both of the categories described by Schiappa in his chapter: the rhetoric *of* mathematics and the rhetoric *in* mathematics. As he explains, "The 'rhetoric *of* mathematics' refers to the deployment of mathematics to enhance the credibility and persuasiveness of a particular argument." What makes the rhetoric of the National Council of Teachers of Mathematics interesting is how it seeks to apply ideas accepted within the mathematics community in other disciplinary contexts. As this chapter will show, the NCTM's argument that mathematics is important across a spectrum of activities serves as a rationale for the educational reforms it believes need to take place.

I would argue, however, that one cannot simply look at the rhetoric *of* mathematics without also at least giving a brief examination of the rhetoric *in* mathematics. In Schiappa's words, "rhetoric *in* mathematics can be understood as the argumentative and stylistic modes of persuasion found in written proofs and arguments throughout the history of mathematics." Part of what drives the conflict within the field of mathematics education over the Common Core is the tension between professional mathematicians and mathematics educators. What is often assumed in mathematics—that a set of basic, stable operations exists and can be taught in a consistent manner—is often the basis of controversy between mathematicians, K-12 educators, parents, and students.

Although the standards of the Common Core were first discussed in November 2007, the Common Core State Standards of Mathematics (CCSSM) did not start taking shape until 2009 and became part of the initial publication of the Common Core standards in June 2010 (Common Core 2016). As soon as the CCSSM was published, four leading mathematics organizations (the Association of Mathematics Teacher Educators, the NCTM, the National Council of Supervisors of Mathematics, and the Association of State Supervisors of Mathematics) announced their support.

Within the rhetoric and mathematics literature, much scholarly attention has focused on the internal discourse within the field of mathematics (Cyphert 1998; Davis and Hersh 1987; Krips, McGuire, and Melia 1995; Reyes 2004; 2014). Not nearly enough attention has been paid within the rhetoric and mathematics literature to how mathematics is taught, particularly in K-12 education. Some scholars who have devoted attention to the issue have detailed the long-running conflict over whether mathematics should be seen as a subject for the elites or for the masses (Schoenfeld 2004). Others have examined educational reform movements within the philosophy of mathematics education and their relationship to how teachers should teach mathematics on a daily basis (Herrera and Owens 2001). Still others have explored mathematical literacy, particularly how students make sense of mathematics (Pugalee 1999; Van Amerom 2003).[1]

It is in the teaching of mathematics where we see the intersection of what Goodnight (1982) called the personal, technical, and public spheres of communication. Examining these spheres simultaneously is vital to understanding how mathematics is taught in school as "an essential forerunner of any mathematician's thought" (Ernest 1998, 260), and because "mathematics is learned through a process of communication" (Seeger 1998, 7). The study of how mathematics is learned is an important adjunct to understanding how the internal rhetoric of mathematics functions. Rather than consider mathematics as practiced by mathematicians—as other scholars have done (Reyes 2004; 2014; Davis and Hersh 1987; Cyphert 1998)—I will consider mathematics as *learned by students*, in particular how teachers and the curriculum they develop help students learn mathematics and how rhetoric inside of mathematics and about mathematics impacts how students learn.

Garland observed that the Common Core standards "don't include lesson plans, or teaching methods, or alternative strategies for when students don't get it" (Garland 2014a). This pedagogical shortcoming has led a variety of educational observers to conclude that the problem with the Common Core has been poor implementation. With the publication of *Principles to Actions* (a document not part of the Common Core standards), the NCTM sought "to fill this gap between the development and adoption of the CCSSM and other standards and the enactment of practices, policies, programs, and actions required for their widespread and successful implementation" (NCTM 2014, 4). This publication marks the first time that a professional organization entered the Common Core controversy by actually describing what should be done at the classroom level, focusing on the implementation of standards and not just their description (AAE 2014). Documents like *Principles to Actions* are therefore important to rhetorical inquiry because they provide insight into the controversy of teaching the Common Core standards.

This chapter investigates the Common Core controversy to gain insight into the tensions between proponents and opponents of current mathematics pedagogy. Because *Principles to Actions* both responds to controversy and helps teachers and other education professionals understand and carry out the Common Core Standards in Mathematics, it will receive some attention in my analysis. To better understand *Principles to Actions*, I will first examine the context in which the text was created. I will then investigate the various philosophies of mathematics from which critiques of the document arise before finally discussing what its case reveals about the rhetorical construction of mathematical knowledge.

A Brief History of Mathematics Education Reform and the NCTM's Role in It

Historically, the views of the general public and the mathematics community about mathematics education have changed as opinions about what students can and should learn about mathematics have evolved (Schoenfeld 2004). Examples of critiques of mathematics education can be found as far back as 1908, when William Betz, a high school teacher from Rochester, New York, related a conversation he had had about mathematics education: "A few months ago, the principal of a well-known school said to me,'the teaching of geometry has become stale. Something must be done to put new life into it'" (Betz 1908, 625). The remainder of the twentieth century and the beginning of the twenty-first century witnessed a variety of movements designed to change how mathematics was taught in grades K-12. As the Cold War heated up with the space race between the United States and the Soviet Union, traditional math was seen as no longer sufficient. With the launch of Sputnik in 1957, a "new math" movement developed that emphasized such topics as set theory, mathematical laws (e.g., the associative, commutative, and transitive properties), and the deductive method (Hayden 1981). The movement failed in large part because it neglected the psychological aspects of education, and because teachers were not given adequate preparation to teach under a new paradigm (Hekimoglu and Sloan 2005).

"New math" led to "basic skills" math, which focused on computation, a skill experts hoped teachers would be better equipped to teach (Abbott et al. 2010, 28–31). However, that too proved not to be the case (Hekimoglu and Sloan 2005). Following the failure of the "basic skills" approach to teaching mathematics, there was a gap during which no specific paradigm informed mathematics education. Instead, there were a variety of reform documents and opinion pieces, most of which could be classified as supporting a "problem-solving" approach to math (Pollak 1983; Rappaport 1971; Beberman 1958; Buck 1968; Woodward 2004; Commission on Mathematics 1959). While well intentioned, many of these reports did not generate much support for remaking the mathematics classroom in their image.

In 1980 the NCTM published a report entitled *An Agenda for Action*, which both encapsulated and expanded previous reform movements in mathematics (NCTM 1980; Ball 1991). *An Agenda for Action* was important because it served as the basis for later reform movements of the NCTM such as the *Standards* (NCTM et al. 1997). Several of its proposals, such as a decreased emphasis on computation, an increased reliance on calculators and computers, and a decreased emphasis on calculus in high schools, found their way into later NCTM documents. These proposals drew a great deal of criticism in the collegiate and professional mathematics communities because of their significant shift from traditional practice (Klein 2003).

The next major inflection point in education reform following *An Agenda for Action* was the publication of *A Nation at Risk*. The report, published by the National Commission on Excellence in Education, highlighted problems in K-12 education and suggested immediate solutions to problems in the school system (Park 2004). *A Nation at Risk* was meant both for public consumption and for policy-makers. The report's opening statement includes the competitive *topos* of <falling behind> which is typical in educational reform documents: "The educational foundations of our society are presently being eroded by a rising tide of mediocrity that threatens our very future as a Nation and a people. What was unimaginable a generation ago has begun to occur—others are matching and surpassing our educational attainments" (Gardner 1983, 1).

Reform rhetoric also involved common appeals to heroic teachers as vital to the success of American education (Steudeman 2014), and claims of dire consequences for American exceptionality in the absence of swift action (Holmes 2012). The report generated continuing discussion about educational reform, even more than thirty years after its initial publication (Ansary 2007; Park 2004). More importantly, it provided impetus and rationale for the future work of the NCTM. To solve the nation's educational problems, *A Nation at Risk* called for increased investment and reform in mathematics education. Two conferences—one commissioned by the National Science Foundation (NSF) and the other jointly sponsored by the NCTM and the University of Wisconsin— generated two reports that ultimately created the NCTM's Commission on Standards for School Mathematics (Romberg 1993). That commission then produced a draft report in 1987 that would later become the *Curriculum and Evaluation Standards for School Mathematics* (NCTM 1989). Over the next several years, the commission published the *Professional Standards* and the *Assessment Standards*. These three volumes of the standards were replaced in 2000 with the publication of *Principles and Standards of School Mathematics* (NCTM 2000). The publication of the various standards led many in the educational community to see the NCTM as a leader in mathematics education reform (Romberg 1993; Briars 2014a; Ball 1991).

Principles to Actions is the sixth volume of the NCTM's work on teaching standards. Rhetorically, it responds to other philosophies of mathematics education that became prominent in the 1980s as a consequence of the publication of *Curriculum and Evaluation Standards*. Before I look at *Principles to Actions* in more detail, the underlying philosophies of mathematics and mathematics education, in particular the question of what constitutes mathematical knowledge, require some focused discussion. As I will show, NCTM reform rhetoric has been grounded in *educational* constructivist (as opposed to sociological) rhetoric.

Dominant Philosophies in Mathematics Education

Mathematics, like rhetoric, has long been concerned with epistemological questions such as What is a mathematical object? and How do we know what a mathematical object is? How one answers these questions determines the rhetorical strategies one believes are the most efficacious for teaching mathematics. As this brief analysis will demonstrate, there are a range of views, from abstract to concrete, on how to answer these questions. The major philosophical positions on mathematics education can be distilled into three overarching viewpoints: formalism/Platonism, fallibilism, and constructivism.[2]

Formalism and Platonism are related philosophies that hold that mathematics is revealed and discovered, rather than created. Formalists deny any truth value to mathematics, insisting that truth presents itself only in logical consequences,[3] while Platonists view "mathematical truth to be necessary—true in all possible worlds—and mathematical knowledge to be a priori and certain" (Maddy 1989, 1124, 1132). Wilder sums up the Platonist view succinctly when he writes, "mathematics exists in a world of ideals, ready and waiting to be discovered by the investigator" (Wilder 1981, 27). Formalism's logical consequences are the means by which Platonism is upheld. As the Latvian mathematician Karlis Podnieks observes, "mathematicians are accustomed to thinking of mathematical structures as independently existing objects" (2015, 3). Many mathematicians, such as Kurt Gödel, work within a Platonist perspective.

Fallibilism is the belief that "mathematics consists of language games with deeply entrenched rules and patterns that are very stable and enduring, but which always remain open to the possibility of change, and in the long term, do change" (Ernest 1999b). For fallibilists, such as the mathematics education researcher Paul Ernest and the philosophers Imre Lakatos and Karl Popper, social forces play an important role in shaping mathematics. Ernest (1999b, n.p.) cites Wittgenstein, another fallibilist, who argued that "we often follow rules in mathematical reasoning because of well-tried custom, not because of logical necessity." One of the reasons that fallibilism has not taken hold in mathematics education is that Platonism is considered more fruitful than fallibilism in "stimulating and sustaining creative work in mathematics" (Lerman 1998, 291). If we doubt every aspect of mathematics and believe it susceptible to failure, then it becomes difficult to establish solid foundations for teaching and building student knowledge.

The third mathematical philosophy is constructivism, the philosophy that underlies the NCTM's work. Unlike other areas of mathematical philosophy, constructivism arises both from philosophy and from cognitive and developmental psychology (Noddings 1990). Much of what passes as constructivism in mathematics education arises from Piaget, who believed that mathematical knowledge was constructed through reflective abstraction. Accordingly, the goal of a constructivist pedagogy is to develop the appropriate cognitive structures to understand mathematics (Noddings 1990). Because of the prominence of Piaget within education as a whole, and within elementary education in the 1970s, constructivism was seen by NCTM as being the ideal approach to teaching mathematics (Copeland 1970). A variety of research has identified the influence of constructivism at the core of the NCTM's reform efforts, beginning with the first reform in the 1990s. Martin Simon observes, "Constructivist perspectives on learning have been central to much of recent empirical and theoretical work in mathematics education and, as a result, have contributed to shaping mathematics reform efforts" (1995, 114). In addition, Lerman notes, "Within the mathematics education community, mathematics can be claimed to be a social construction and each person is seen to (re-)create that mathematics for her/himself" (1998, 295). Lerman also notes that constructivism in a mathematical sense is based on "the right of all learners to be seen to be creating their own mathematics for themselves, and countering a view of the teacher as authority in relation to knowledge" (1998, 294).

Various educational philosophies define mathematical knowledge differently as well as argue that mathematical knowledge is acquired in different ways. Ernest, a fallibilist, suggests that mathematical knowledge has five components: a language, a set of accepted statements, a set of accepted reasonings, a set of questions selected as important, and a set of metamathematical views, which include standards for proof and claims about the scope and structure of mathematics (Ernest 1998, 249). For the most part, each of the three major philosophical positions outlined earlier would concur with Ernest's components, though they would have philosophical differences about how to characterize the first and last ones. In this analysis, I will focus on the first component: what it means to communicate mathematically. Among teachers, students, and parents, there has been a tendency to describe mathematics as a special language, not unlike a foreign language (Kenney 2005). Language use among mathematicians does require a shared understanding of what various symbols mean (Davis and Hersh 1987). Once acquired, "the specialized language in which mathematicians converse with each other is a magnificent tool for conveying complex ideas precisely and swiftly" (Ellenberg 2014, 12). In addition to a shared technical vocabulary, mathematical communication requires an agreed-upon system of proving. In algebra and geometry classrooms, for example, mathematics is often taught as a system of proofs with particular structures (e.g., the two-column proof) where the deductive logic between each step is of primary significance. As Ernest notes, this practice has come under pedagogical scrutiny: "Although traditionally it has been thought that the acceptance of mathematical knowledge depends on having a logically correct proof, there is growing recognition that proofs do not follow the explicit rules of mathematical logic, and that acceptance is instead a fundamentally social act" (1999a, 73).

The various philosophical viewpoints mentioned earlier create questions about how mathematical knowledge is socially accepted and acknowledged. While there are exceptions, many mathematicians tend to fall into the formalist view, sociologists of mathematics education tend toward fallibilism, and much of the NCTM's research is based on constructivism. My analysis follows Ernest's fallibilist conception, which is implicit within the rhetoric of mathematics (Reyes 2004; 2014). As a contemporary illustration of Ernest's point about how mathematical knowledge is created and accepted, consider the teaching of high school geometry. Many of us learned geometry using the two-column proof. As Falk Seeger explains, "Although many people believe that deductive proof helps define the essence of mathematical reasoning, proofs have been disappearing over the last decade or so from their traditional home, the course in Euclidean geometry" (1998, 12). This shift mimics the change that has taken place within the discipline of mathematics itself from having only structured proofs to having a variety of different forms of proof.

Principles to Actions as a Response

With different philosophies of mathematics education, there are varying views about what should be taught to students and when. These divergent views have led many, particularly within the general public, to wonder whether standards for teaching mathematics have been lowered, or whether they have become irrelevant (Romberg 2017). Within the mathematics community, however, there is a very strong belief in the existence of clear standards. As Jason Zimba, one of the coauthors of the Common Core, argues, "There's actually very little fuzziness to the math in the Common Core. Students have to memorize their times tables by third grade" (qtd. in Garland 2014a). If the standards in the Common Core are so clearly defined, however, why have there been so many complaints by teachers, administrators, and the general public about the Common Core, particularly in mathematics? The Common Core Standards do not tell a district or a teacher how they should be implemented and do not include lesson plans, strategies for teaching, or ways to help students who do not understand a particular concept learn it (Garland 2014a).

NCTM's sixth volume, *Principles to Actions*, was published in 2014 after the Common Core standards had been released, as an effort to bridge the gap between mathematical experts and teachers, administrators, and parents (2014, 4). In its preface, the document identifies these groups as the primary audience for the text: "*Principles to Actions* is for teachers, coaches, specialists, principals, and other school leaders. It is for policy-makers and leaders in districts and states, including commissioners, superintendents, and other central office administrators. Moreover, it will give families guidance about what to look for and expect in the system educating their children" (NCTM 2014, vii–viii).

As the remainder of this chapter will show, *Principles to Actions* is a document shaped by its place in the educational reform movement, constrained by its adherence to the constructivist paradigm, and influenced by a rhetorical approach to learning mathematics. To support this characterization, I will draw on Barry Brummett's notion of the representative anecdote.

Principles to Actions as Representative Anecdote

Barry Brummett's concept of the representative anecdote provides a framework for understanding the rhetoric of the NCTM. Brummett's approach allows the analyst to represent "the essence of discourse by viewing it as if it follows a dramatic plot" (1984a, 3). Anecdotes necessarily involve as a part of their plot the declaration of good and bad (Brummett 1984a; 1984b). Since the NCTM is trying to establish *why* previous ways of approaching mathematics education are ineffective, they must advance a narrative that identifies both the problems in the previous practices and how the changes both inside and outside the mathematics classroom can improve educational outcomes.

An assessment of *Principles to Actions*, as well as the antecedent work created by the NCTM, reveals three representative anecdotes. First, the current state of mathematics education is good, but not great. Making it great again requires an update rather than a repudiation of previous standards. Second, mathematics pedagogy is inherently unequal, particularly when it comes to race and gender. Steps must be taken to solve equity issues. Finally, the NCTM is the authority on what is best when it comes to mathematics education. Each of these anecdotes is a necessary constituent of NCTM's problem-cause-solution approach. What makes *Principles to Actions* interesting is that each of the anecdotes is connected both to the global disciplinary level and to the local level of the individual teacher. It is important to note that the representative anecdote itself is not a story. Rather, the anecdote is a short summation of a narrative that has been designed to influence change. In the case of *Principles to Actions*, the anecdotes utilize a format that has existed ever since the time of the ancient Greeks in classical rhetoric. What makes these anecdotes interesting for the rhetorical critic is that the NCTM is the protagonist at the heart of each anecdote; the solution can only come from the NCTM. As such, these anecdotes constitute a rhetoric *of* and *in* mathematics. Each of these anecdotes will be considered in the analyses that follow.

Anecdote #1: Math Education Is Good but Not Great

The first anecdote—that the current state of mathematics education is good, but not great, and requires an update but not a repudiation of previous standards to return to greatness—seems inherently contradictory. Why would an organization contend that its previous work has not solved the problems it has identified? The NCTM utilizes this anecdote to embed itself further into the system of mathematics education. As the NCTM notes in the beginning of *Principles to Actions*, "the revisions to this updated set of Principles reflect more than a decade of experience and new research evidence about excellent mathematics programs, as well as significant obstacles and unproductive beliefs that continue to compromise progress" (2014, 4). The text notes in the section on professionalism that "a professional does not accept the status quo, even when it is reasonably good, and continually seeks to learn and grow" (NCTM 2014, 99). The NCTM tries to balance here several different considerations. In the problem-cause-solution paradigm, the NCTM could be considered part of the problem because it has provided standards, but not advice, on how to achieve those standards. This also may explain why NCTM positions *Principles to Actions* as an extension of their previous work and not a revision. If they had created Standards 3.0, many in the potential audiences might have raised questions about what was wrong with the previous standards.

In the first and final chapters, the key takeaway for teachers is that they should continue to be lifelong learners and that they should collaborate more with their colleagues. This in and of itself is not terribly interesting except when it becomes part of the expectations placed upon teachers later in the text. Leaders and policymakers are given a sixteen-item list of suggested actions. Coaches and principals are given twenty-one items to consider. Teachers, however, are given twenty-seven different items that they should consider or change in their teaching. Many of these items are connected to the actions of school administrators and political leaders. One example of such a teaching goal is for teachers to "understand and use the social contexts, cultural backgrounds, and identities of students as resources to foster access, motivate students to learn more mathematics, and engage student interest" (NCTM 2014, 115). As we will see later in the chapter, the dependencies of teaching, administrative, and legislative goals create obstacles for teachers endeavoring to carry out this expectation.

The other challenge arising from NCTM's good-but-not-great anecdote is the role of test scores in demonstrating a need for *Principles to Actions*. If scores are decreasing, then the decline can be attributed to existing standards created by the NCTM. If the scores are increasing, then there is less of an inherent need for a new volume of standards. In the introduction, the NCTM tries to preserve the credibility of its program while at the same time make a case for the need for improvements. To protect their credibility, they indicate that SAT, ACT, and National Assessment of Educational Progress scores have increased, as have the number of students both taking and passing Advanced Placement calculus and statistics exams. They attribute these successes "in large measure to the leadership of NCTM" (NCTM 2014, 1). They also note, however, areas of concern, including the number of students deemed ready for college mathematics and results of the Program for International Student Assessment (PISA) exams.

Another strategy the NCTM uses to escape their own dilemma is to argue that it is not necessarily the Common Core standards but their implementation that is responsible for educational shortfalls. The organization argues that "effective teaching is the nonnegotiable core that ensures that all students learn mathematics at high levels and that such teaching requires a range of actions at the state or provincial, district, school, and classroom levels" (NCTM 2014, 4). As this analysis will later show, individual teachers and parents are seen as a stumbling block to further improvements in children's mathematical achievement.

Because the NCTM chose to update rather than revise previous standards, little changes in its philosophical approach to mathematical education. Like all of the volumes before it, *Principles to Actions* is based on a very strong constructivist paradigm (Klein 2003). The continued unquestioning adherence to this paradigm is the point of contention for many of the NCTM's critics. Seeger, for example, writes, "For many students, using representations and manipulatives as a remedy to learning problems is of little help. For them, it actually creates a new problem, because these students are now facing two problems: to understand the intention of representations and to understand the world of numbers without being able to make a connection between the two" (1998, 309). For Seeger the constructivist paradigm impedes students' capacity to connect abstract mathematical language and symbols to the objects and relationships they are supposed to represent. Steinbring also observes that "the core problem of understanding school mathematics [is that] . . . students have to decipher signs and symbols!," which presumably is made harder by the constructivist approach (1998, 345).

Anecdote #2: Mathematics Teaching Is Unequal and Requires Equity

The second anecdote is that the current state of mathematics teaching is inherently unequal, particularly for students from disadvantaged backgrounds. Many scholars and parents believe that there are inequalities in the present system of mathematics teaching that the NCTM fails to acknowledge. Their guiding principle (one of the six for *Principles to Standards*) for Access and Equity is that "an excellent mathematics program requires that all students have access to a high-quality mathematics curriculum, effective teaching and learning, high expectations, and the support and resources needed to maximize their learning potential" (NCTM 2014, 5).

The Diversity in Mathematics Education Center noted that many of the explanations of differences in mathematics ability reported by the media centered on differences across broad groups of students. They explained that these differences ultimately create a vicious cycle: problems with learning math are attributed to being disadvantaged, and this label reinforces achievement differences between groups, creating further difficulties for those already identified as being disadvantaged. For example, the mathematics education of African American students, they argue, "is shaped by a master narrative within society that positions African Americans as less capable in mathematics than their White peers" (Diversity 2007, 409).

The NCTM does recognize equity concerns in *Principles to Actions* and takes a broad view on the kinds of inequities that might influence education. The organization comments:

> Attending to access and equity also means recognizing that mathematics programs that have served some groups of students, in effect privileging some students over others, must be critically examined and enhanced, if needed, to ensure that they meet the needs of all students. That is, they must serve students who are black, Latino/a, American Indian, or members of other minorities, as well as those who are considered to be white; students who are female as well as those who are male; students of poverty as well as those of wealth; students who are English language learners as well as those for whom English is their first language; students who have not been successful in school and in mathematics as well as those who have succeeded; and students whose parents have had limited access to educational opportunities as well as those whose parents have had ample educational opportunities. (NCTM 2014, 60)

Each of these statements serve as a response to previous criticisms of mathematics education, or to particular stereotypes about various groups, and thus their inclusion in the equity statement was intentional. Each of the groups mentioned in the NCTM's equity statement has been stereotyped as being unable to succeed in mathematics. Two examples of stereotypes about mathematics that were mentioned above have to do with women and Native Americans. One of the stereotypes that exists about women in mathematics is that they are less capable of doing well in the subject than men are (Tobias 1978; Tobias and Weissbrod 1980). The NCTM endeavors to counter this stereotype. Danica McKellar, who is discussed in chapter 8 by Wynn, for example, has been cited by former NCTM president Skip Fennell as a "terrific role model" (*Newsweek* 2007). The NCTM also raises issues of inequality in the mathematics educations of Native Americans, fueled by stereotypes like the ones G. Donald Allen, a mathematician from Texas A&M, describes: "For the American Indians north of Mexico, we may say that although their bonds of superstition and lack of an adequate number symbolism limited their mathematical progress, number still played an important role in their religious beliefs" (n.d., n.p.).

The NCTM's equity statement referenced above does address potentially disadvantaged groups based on factors such as socioeconomic status, gender, and their English language competency, but only on a theoretical level. The problem with the equity statement, as well as the rest of the chapter, however, is that it does not explain precisely *how* these inequalities should be reconciled within the classroom. *Principles to Actions* does include case studies at the end of each chapter to demonstrate how equity can be fostered in the classroom. Most of the case studies utilize teacher and/or student dialogue and take place in a classroom setting. The case study in the equity chapter, for example, deals with how curriculum might be retooled to create a new course to better serve students at or below grade level. It does not, however, address how to remedy the inequality through student-teacher interactions.

The NCTM is caught between two contradictory perspectives in their efforts to promote equity. On one side is the traditional impulse toward formalism, teaching "through memorizing facts, formulas, and procedures and then practicing skills over and over again. This view perpetuates the traditional lesson paradigm that features review, demonstration, and practice that is still pervasive in many classrooms" (NCTM 2014, 9). On the other, there is an impulse to adopt an ethnomathematical perspective. Ethnomathematics is a relatively new field in the study of mathematics. The term was first coined in 1977 by the Brazilian mathematician Ubiratan D'Ambrosio (Izmirli 2011). More recently, Abbott et al. define ethnomathematics as an approach that "emphasizes learning the mathematical traditions of all nations, including the early calendars and mathematical ideas of civilizations long gone." From a pedagogical perspective, the standard curriculum for ethnomathematicians is "not the birthright of all people but rather a form of colonialism" (2010, 31).

The problem with *Principles to Actions* from an ethnomathematical perspective is that it views the actual content of mathematics as culture free. New research by scholars of color challenges this perspective. Yvette Solomon, an educational researcher at Lancaster University, for example, argues that mathematics education under the current paradigm supported by the NCTM creates a "damaging view among teachers that mathematics is culture-free and universal, and hence not subject to potential inequities, being only predicated on 'ability'" (2009, 145). Danny Bernard Martin, a professor of mathematics and the chair of Curriculum and Instruction at the University of Illinois at Chicago, pushes Solomon's critique farther, taking direct aim at the NCTM and *Principles to Actions*. He writes, "The unspoken, hidden reality in *Principles to Actions* is that potential benefits to the collective Black are metered by Whites and White design and are contingent on parallel benefits to Whites. *Principles to Actions* could never have been written to focus solely on gains for the collective Black. . . . The hard truth is that the outcomes and inequities lamented over in *Principles to Actions* and previous documents are precisely the outcomes that our educational system is designed to produce" (2015, 21).

While ethnomathematics is not yet dominant in the field of mathematics education, it represents an approach that the NCTM must ultimately respond to, because many scholars and teachers of color are pressing the organization to recognize mathematical logic and history from different racial and historical traditions as a way of demonstrating equity. Maintaining the position that culture influences our understanding of mathematics, however, places the NCTM at odds with its own traditional formalist stance.

Anecdote #3: NCTM Knows Best

The third representative anecdote in *Principles to Actions* is that the NCTM knows what's best for K-12 students of mathematics. In examining this anecdote, I will look at this idea from the NCTM's perspective, which is that it knows best how to teach mathematics and others do not, as well as from the critical external position that the NCTM does not know what is best for reform and, arguably, cannot be an agent of change. Interestingly, both of these perspectives point to the failure of tradition as support for their position.

Given the problems cited both in achievement and in equity, the NCTM must demonstrate through *Principles to Actions* that they are best for solving the problems of mathematics education. Their challenge is, of course, how to make the case that they are the solution rather than the cause of the problem given their historical involvement in developing the policies and practices of the currently problematic system. Kenneth Burke's concept of victimage helps explain how they make this case. Burke notes that when man, who is goaded by hierarchy and rotten with perfection, fails to be perfect, then guilt results. One of the strategies for purging guilt is through victimage, which requires a scapegoat who is blamed for the social imperfections that exist (Burke 1970, 1984). In the case of the NCTM, there are three scapegoats: parents who want to see old ways of teaching continue in their child's classroom, teachers who fail to properly carry out the directives of the NCTM, and math professionals who do not agree with the organization's constructivist viewpoint. Each of these groups either unwittingly (in the case of parents, most of whom do not know the NCTM exists) or wittingly (in the case of teachers and math professionals) is responsible for poor practices in mathematics education because of its conservative adherence to traditional viewpoints.

The NCTM has struggled with how to respond to two questions: Is there one right answer to a problem? And how is mathematical knowledge generated? Chazan elucidates the traditional perception of parents and other educators about mathematics: "The mathematics of the teacher or textbook is not contested, or contestable, knowledge . . . the teacher actively tries to clear up confusions as quickly as possible" (2000, 117). For many parents, Chazan's words describe how they learned mathematics that echoes the formalist paradigm. The NCTM, however, relies on a constructivist paradigm in its philosophy of math education (Germain-McCarthy 1999). Under a constructivist paradigm, students are responsible for cognitive reflection as they construct their own knowledge of the world about them. The National Science Foundation summarized the NCTM's position on constructivism: "Knowledge is socially constructed through human activity, shaped by context and purposes, and validated through a process of negotiation within a community of practice. Thus, it is always tentative rather than absolute" (Borasi and Fonzi 2002, 14). The implication of the NCTM's constructivist philosophy, from a parental perspective, can "smack of relativism, namely, that there are no solid grounds for preferring certain conceptions over others, that everything, including mathematics, is relative," which would be "the height of pedagogical irresponsibility" (Chazan 2000,

117). Part of the problem Chazan identifies is that parents believe that if there is one right answer, then there is one *best* way to get that answer. The constructivist philosophy challenges that assumption by arguing that students can often find ways of solving problems that, while not "conventional," still arrive at an appropriate answer. As a result, when a teacher demonstrates a different way of solving problems, or when students generate a longer but still accurate solution, a large subset of parents still believe the teacher is *wrong* for allowing that kind of discussion in the classroom (Briars 2014a; 2014b). Research by Roberts about parents and the CCSSM suggests that many parents still believe that students "need to learn to just quickly get the correct answer" and that some of the "Common Core things I have seen are not logical ways to get the answer" (2015, 42).

NCTM's constructivist approach also deals with how students learn algorithms. One illustration of how the organization tried to change the perception of algorithms appears in the section on teaching and learning. When teaching multidigit multiplication, *Principles to Actions* encourages the creation of an "accessible" algorithm (the example on the left) as opposed to the "conventional" algorithm, which involves carrying digits (the example on the right) (NCTM 2014, 45).

$$
\begin{array}{r}
46 = 40 + 6 \\
\times\, 68 = \underline{60 + 8} \\
2400 = 60 \times 40 \\
360 = 60 \times 6 \\
320 = 8 \times 40 \\
\underline{\quad 48} = 8 \times 6 \\
3128
\end{array}
\qquad
\begin{array}{r}
46 \\
\underline{\times\, 68} \\
368 \\
\underline{2760} \\
3128
\end{array}
$$

NCTM President Diane Briars spoke of the differences between these approaches, and what teachers needed to do to help parents understand the accessible approach, calling it "a critical priority for your beginning-of-the-year engagement efforts" (2014a, n.p.) She explained that "because of their own school experiences, many parents hold beliefs about teaching and learning that *Principles to Actions* describes as 'unproductive'" (2014a, n.p.). The problem with the NCTM's stance is that parents want to help their children but feel disenfranchised and helpless because the Common Core and the NCTM *Standards* do not reflect their own methods for and experiences with doing mathematics. NCTM's new standards communicate to parents that their way of learning is no longer appropriate for their children, and thus they must change. Briars's rhetoric is intended to convince teachers that they and the NCTM know what is best for children.

Though parents receive the brunt of the NCTM's critique in *Principles to Actions*, teachers are also made scapegoats for the failures in mathematics education. As the report explains, "mathematics educators must hold themselves, individually and collectively, accountable for all students' learning" (NCTM 2014, 99). In their discussion of teachers' responsibility to students, the NCTM identifies the improper pedagogical perspectives that obstruct learning goals. The NCTM observes, for example, that though parents may "unwittingly reinforce the notion" that some students cannot do well in mathematics by "excusing low performance by their children as genetic destiny," teachers "reinforce this misconception by sorting students by ability, believing that some [students] can 'do math' and others cannot" (NCTM 2014, 62). Further, the NCTM chastises teachers for clinging to outdated beliefs about mathematics teaching: "Many parents and educators believe that students should be taught as they were taught, through memorizing facts, formulas, and procedures and then practicing skills over and over again" (NCTM 2014, 9). The NCTM is quick to correct teachers, but there are constraints in the classroom that make it difficult for educators to adopt new standards. One of these constraints is the amount of time teachers have to cover material in their course. Chazan writes, "there is always pressure to 'cover' the material. There is so much for students to know. Even if one would like students to be more active in class, because of time constraints, one cannot reasonably expect students in school to blunder inefficiently as they attempt to recreate mathematics it has taken generations of talented mathematicians to create" (2000, 117). Because of these constraints, teachers often stick with strategies for doing mathematics that they had been taught (NCTM 2014, 10). *Principles to Actions* suggests that teachers should significantly modify their classroom routines and methods of discussion to increase the use of student-led discourse. However, the organization also recognizes that teachers need help to do so. *Principles to Actions* highlights several different approaches to helping teachers, such as the use of coaches and specialists, extra time for mathematics instruction, as well as collaborative teacher preparation and assessment of classroom instruction. Despite recognizing these pedagogical challenges, the NCTM still insists that teachers bear substantial responsibility for the current shortcomings in mathematics education.

The final scapegoat in NCTM's critique is mathematics professionals, particularly in colleges and universities, who do not share the NCTM's philosophy of constructivism. The adoption of constructivism as a philosophical foundation for pedagogy is a point of great contention within the mathematics community. Constructivist rhetoric can be seen as a form of pluralism, particularly when it comes to describing mathematics (Hellman and Bell 2006). Many mathematics professionals reject a constructivist approach, arguing instead for a Platonic view (Klein 2003; Balaguer 2009). Though mathematicians participated in the creation of the CCSSM and the revision of the Standards, the Mathematical Association of America (MAA)—a group consisting largely of mathematicians and mathematics professors—listed nine different concerns about the *Principles and Standards for School Mathematics*, several of which remained largely unaddressed in *Principles to Actions* (Briars 2014a; Ross 2000). Mathematics professionals critiqued NCTM's curriculum for lacking basic skills, pure mathematics, prevision in mathematical discourse, and geometry based on a "logical sequence of theorems based on axioms" (Ross 2000). Some more traditional members of the MAA, along with parents and teachers, rejected the anecdote <NCTM knows best>. In response, the NCTM has tried to marginalize those who reject its ideals. One of the ways in which the NCTM has marginalized its critics is by withholding accreditation according to the Council of Accreditation of Teacher Preparation (CAEP).[4] In other words, if a bachelor's or master's program in mathematics education does not teach according to the NCTM's *Standards*, the program cannot be accredited. Given that the Mathematical Association of America (MAA) now encourages changes that encourage college mathematics programs to engage in less lecture, more active learning, and more application in a variety of mathematics courses,[5] the NCTM has further power to scapegoat those who do not align with their philosophies.

Implications and Conclusion

This analysis sheds light on the dual nature of how we perceive mathematical knowledge. One viewpoint is to believe that mathematics is more certain than other types of knowledge (Chazan 2000, 131; Rotman 1993, 2000). As this analysis has shown, attacks on the certainty of mathematical knowledge have come from critical theory, from fallibilism, and even from within the field of mathematics itself, as *Principles to Action* and the rhetoric of the NCTM attest.

While *Principles to Action* was able to persuade a variety of audiences that mathematics is less certain than they believe, there are still some audiences that were not fully persuaded by its rhetoric. I believe there are two significant threads worth pursuing in future research. The first has to do with how race and critical theory have become a part of the rhetoric of mathematics pedagogy, and the second concerns the role that the authority of professional organizations such as the NCTM plays in disciplinary and cross-disciplinary discussions.

First, this analysis suggests further ways the role of race and culture needs to be considered when thinking about mathematics education. It is not enough to acknowledge racial inequity and move on. Previous analyses by mathematics and communication scholars in the rhetoric of mathematics have come primarily from a perspective that has separated culture from the actual nature of mathematics (Davis and Hersh 1987; Reyes 2004; 2014). While scholars in technical communication and other portions of the rhetoric of inquiry have begun to investigate intersections of race, rhetoric, and pedagogy (Haas 2012), rhetoric and mathematics research has not yet done so. Mathematics educator Danny Martin's project, which critiques the NCTM, can serve as a starting point for an approach that more closely ties a study of race to mathematical pedagogy. Martin recognizes in the NCTM that when it comes to education "practice, research, and policy—[are] predominantly White." He reminds us that "about 85% of the US teaching force is White" and "that the research and policy domains are also characterized by a largely White demographic" (Martin 2015; see also Martin 2013; 2008; Martin, Gholson, and Leonard 2011). In Martin's estimation, more investigation needs to be done into why African American students face challenges in mathematics.

This viewpoint has been absent from previous research in the mathematics education community and could serve as a starting point for other critical projects within the rhetoric and mathematics. What rhetorical scholars can bring to the conversation is an understanding of both the micro- and macrolevels of representation and persuasion. Ethnomathematicians have focused on critiquing the notion that the mathematics classroom is a politics-free zone. Rochelle Gutierrez, for example, imagines a kind of critical mathematics in which students both recognize and confront sites of inequity and utilize mathematics to help address those situations. Brian Katz, an algebraic geometry scholar, also notes that "math is political because it is done by groups of people. Those people construct the narratives of our discipline, and telling any story involves selecting characters and emphasizing a plot; the one I learned is almost exclusively about European men" (2017, n.p.).

The exploration of ethnomathematics would add a new subfield to the study of rhetoric and mathematics—rhetoric and mathematics education—whose exploration would help us further understand how children and adults make sense of mathematics, and how the ways in which people make sense of mathematics are reaffirmed or rejected in the educational system. Ethnomathematics can help to further expand our understanding of what is mathematics, and how systems of mathematics have been constructed. By implication, if certain ways of thinking mathematically are reified, then impacts may be experienced from the classroom to the acceptance or rejection of certain kinds of arguments in professional mathematics.

The second implication of this chapter has to do with the NCTM and its attempts to influence the broader policy and cultural debate in the United States about the role of mathematics in society and, ultimately, the utility of mathematics. This analysis highlights the NCTM's problems and explores the ways it might mitigate the conflicts in which it finds itself. The problems include how to change the mindset of parents and teachers both within the discipline and in the classroom. As such, this research serves as a response to the call of Lyne and Howe, who encouraged scholars in the rhetoric of science to examine both how particular discourses represent themselves and how they implicitly and explicitly present their opposition (1986, 144).

If indeed it is difficult to change the perception of how mathematics operates within culture, it will be hard for the NCTM to enact changes in professional practice. From an epistemological standpoint, as Lerman noted, there are potential contradictions in the educational constructivist position that underlies the NCTM Standards: "If the mathematical knowledge that we have is a social construction, but individuals construct their own knowledge, how does that intersubjective social knowledge become intrasubjective?" (Lahman and Lietzenmayer 2015, 288). As Chazan succinctly inquires, "If the teacher knows what is right and what is wrong, what is there to discuss?" (2000, 115). The NCTM's response to Chazan would be, "How do we encourage students to articulate *how* they know what they know? And how do students convince each other and themselves that their answer is correct?"

NCTM's desire for more and longer problem-solving sessions has unwittingly placed them within a larger social conflict over how we as a society view elementary education. Over the last decade, as a result of Common Core and other education-reform movements, there have been increases in standards and more rigorous curriculum at younger ages. Part of the problem that the NCTM faces is that what they perceive to be more challenging instruction is not considered rigorous by some members of the education establishment and the public: "Parents expect their children to be taught 'good' mathematics, which for most of them is what they were taught in school; the society has its own expectations of what is required, but even where there is the recognition that new knowledge is needed, this is to be added without deleting any of the 'basic' [mathematical] content" (Van der Blij, Hilding, and Weinzweig 1980, 46–47).

As this analysis has shown, part of the problem society faces is that there are multiple conceptions of what mathematics, and, more importantly, its teaching, should be about. As noted earlier in the chapter, part of the reason that the NCTM's stance has failed is that parents and other members of the public are used to and desire traditional mathematics teaching, while the NCTM itself has adopted a reform-based paradigm.

Arguably, part of the reason the NCTM's standards are not seen as rigorous by the public has to do with the textbooks and curriculum that are being used. Garland (2014a) noted that many of the textbooks at the time the Common Core came out were merely repackaged versions of what the textbook publishers were already using, rather than a significant change from what had taken place before. Suitable curriculum and textbooks do not exist in large quantities, which impacts teachers who are starting out in education and trying to uphold NCTM's guidelines (Garland 2014a; 2014b). Thus the NCTM finds itself in a dilemma from which it cannot easily extricate itself, because it is being blamed for something that it has, in part, no direct control over.

Because the NCTM has aligned itself so closely with the Common Core standards, it is also seen as part of the problem when it comes to the actual teaching that takes place in the classroom. As a consequence, the NCTM has to defend itself on both ends of the theory-practice continuum (Copur-Gencturk 2015; Herbel-Eisenmann et al. 2016). The problem the NCTM faces in the realm of practice is time. Teachers who support the reforms of the NCTM have to devote several years to modify their teaching to develop the kind of mathematical expertise necessary to teach under the paradigm of *Principles to Actions* (Copur-Gencturk 2015). As a result, there is not a critical mass of the kind of classrooms the NCTM can point to as exemplars of its desired approach. The relatively small number of teachers and classrooms fully following NCTM best practices results in the NCTM having to utilize the representative anecdote <Trust us. We know what's best> without having the sufficient backing to be truly persuasive.

The NCTM also faces the problem that their work is not well known or well understood outside of the mathematics education community. As Herbel-Eisenmann et al. observe, part of the problem the NCTM faces is that it does not control its own storyline; other politicians and groups have done so (Herbel-Eisenmann et al. 2016). Though only briefly recognized by Herbel-Eisenmann and her colleagues, public opinion also plays a significant role in the kind of story NCTM can tell about itself and its efforts. Many Americans believe, for example, that there are gender-based differences in the ability to learn mathematics; they also believe that people either innately have mathematical ability or they do not (Tobias 1978; Tobias and Weissbrod 1980; Paulos 1988; Green 2014). As this chapter discussed earlier, as long as Americans believe that mathematical ability is inherent rather than developed ("the math gene"), the NCTM faces a formidable obstacle in getting the public to accept their story about mathematics education. NCTM has presented a narrative in which *all* students can do well in mathematics. To be persuasive, however, this inspirational rhetoric needs to be accompanied by concrete examples of students (or teachers) who have profoundly changed their attitudes toward mathematics. Because *Principles to Actions* is so new, there is not much data to support the contentions the NCTM makes in that volume. For it to be persuasive, more evidence is necessary if parents and students are to believe that they can succeed in mathematics.

Finally, additional research should be conducted on other groups who are seeking to change how mathematics is taught. The Association of Mathematics Teacher Educators (AMTE), for example—a collective of college professors involved with mathematics education in pre-K to doctoral programs—has produced standards for doctoral programs in education, such as the *Standards for the Preparation of Teachers of Mathematics* (AMTE 2017). An analysis of the AMTE's rhetoric would be a valuable next step in understanding the rhetoric of how mathematics is taught, both because there is some cross-membership between the NCTM and AMTE and because the AMTE created guidelines for future mathematics educators.

The NCTM plays an important role in how mathematics is taught in the United States. Because the NCTM seeks to inform policy and practice, the organization and its rhetoric are important for scholars interested in rhetoric and mathematics to examine in more detail and can provide a productive site for exploring the interface of society, culture, education, and mathematics.

Notes

1. Though Barbara van Amerom's (2003) study considered how students develop algebraic language and Pugalee's (1999) approach asked what it means to be mathematically literate, both approaches adopted a student-oriented, as opposed to teacher-oriented, approach to studying literacy.

2. I recognize here that philosophers of mathematics tend to use three different categories: intuitionism, formalism, and logicism, categorizing constructivism as a subset of intuitionism. Also, strict philosophers of mathematics would see formalism as a separate branch of mathematical philosophy (Maddy 1989). Because the chapter focuses on philosophies of mathematics education rather than philosophies of mathematics, I have chosen these categories.

3. I recognize that this is a simplistic view. Hilbert, Russell, and Putnam are three philosophers who examine formalism from a subtler perspective.

4. See http://www.nctm.org/Standards-and-Positions/caep-Standards.

5. See "A Common Vision for Undergraduate Mathematical Sciences Programs in 2025" (Mathematical Association of America, 2015).

References

AAE (American Association of Educators). 2014. "NCTM Calls for Action on Common Core Implementation." Last modified April 16. http://www.aaeteachers.org/index.php/blog/1291-nctm-calls-for-action-on-common-core-implementation.

Abbott, Martin, Duane Baker, Karen Smith, and Thomas Trzyna. 2010. *Winning the Math Wars: No Teacher Left Behind*. Seattle: University of Washington Press.

Allen, G. Donald. N.d. "Mathematics Used by American Indians North of Mexico." http://www.math.tamu.edu/~dallen/history/american/american.html.

AMTE (Association of Mathematics Teacher Educators). 2017. *Standards for Preparing Teachers of Mathematics*. https://amte.net/standards.

Ansary, Tamim. 2007. "Education at Risk: Fallout from a Flawed Report." Last modified 2007. https://www.edutopia.org/landmark-education-report-nation-risk.

Balaguer, Mark. 2009. "Realism and Anti-Realism in Mathematics." In *Handbook of the Philosophy of Science: Philosophy of Mathematics*, edited by Andrew D. Irvine, 35–101. Amsterdam: Elsevier.

Ball, Deborah Loewenberg. 1991. "Implementing the NCTM Standards: Hopes and Hurdles." Paper presented at Telecommunications as a Tool for Educational Reform: Implementing the NCTM Standards, Aspen Institute, Queenstown, MD, December 2–3.

Beberman, Max. 1958. "An Emerging Program of Secondary School Mathematics." In *New Curricula*, edited by Robert W. Heath, 9–34. Cambridge, MA: Harvard University Press.

Betz, William. 1908. "The Teaching of Geometry in Its Relation to the Present Educational Trend." *School Science and Mathematics* 8:625–33.

Borazi, Raffaella, and Judith Fonzi. 2002. *Professional Development That Supports School Mathematics Reform*. Foundations 3. Arlington, VA: Division of Elementary, Secondary, and Informal Education, Directorate for Education and Human Resources, National Science Foundation. https://www.nsf.gov/pubs/2002/nsf02084/nsf02084.pdf.

Briars, Diane J. 2014a. "Back to School: The Time to Engage Parents and Families." August 2014. http://www.nctm.org/News-and-Calendar/Messages-from-the-President/Ar chive/Diane-Briars/Back-to-School_-The-Time-to-Engage-Parents-and-Families.

———. 2014b. "Core Truths." March 6. http://www.nctm.org/News-and-Calendar /Messages-from-the-President/Archive/Diane-Briars/Core-Truths.

Brummett, Barry. 1984a. "Burke's Representative Anecdote as a Method in Media Criticism." *Critical Studies in Mass Communication* 1 (2): 161–76.

———. 1984b. "The Representative Anecdote as a Burkean Method, Applied to Evangelical Rhetoric." *Southern Speech Communication Journal* 50 (1): 1–23.

Buck, Charles. 1968. "What Should High School Geometry Be?" *Mathematics Teacher* 61:466–71.

Burke, Kenneth. 1970. *The Rhetoric of Religion: Studies in Logology.* Berkeley: University of California Press.

———. 1984. *Permanence and Change: An Anatomy of Purpose.* 3rd ed. Berkeley: University of California Press.

Chazan, Daniel. 2000. *Beyond Formulas in Mathematics and Teaching: Dynamics of the High School Algebra Classroom.* New York: Teachers College Press.

Commission on Mathematics. 1959. *Program for College Preparatory Mathematics.* New York: Commission on Mathematics.

Common Core State Standards Initiative. N.d. "Grade 2: Operations and Algebraic Thinking—Add and Subtract Within 20." http://www.corestandards.org/Math /Content/2/OA/B/2.

Copeland, Richard W. 1970. *How Children Learn Mathematics: Teaching Implications of Piaget's Research.* New York: Macmillan.

Copur-Gencturk, Yasemin. 2015. "The Effects of Changes in Mathematical Knowledge on Teaching: A Longitudinal Study of Teachers' Knowledge and Instruction." *Journal for Research in Mathematics Education* 46:280–330.

Cyphert, Dale. 1998. "Strategic Use of the Unsayable: Paradox as Argument in Gödel's Theorem." *Quarterly Journal of Speech* 84:80–93.

Davis, Philip J., and Reuben Hersh. 1987. "Rhetoric in Mathematics." In *The Rhetoric of Human Sciences: Language and Argument in Scholarship and Public Affairs,* edited by John S. Nelson, Allan Megill, and Donald N. McCloskey, 53–68. Madison: University of Wisconsin Press.

Diversity in Mathematics Education Center for Learning and Teaching. 2007. "Culture, Race, Power, and Mathematics Education." In *Second Handbook of Research on Mathematics Teaching and Learning,* edited by Frank K. Lester Jr., 405–34. Reston, VA: National Council of Teachers of Mathematics.

Ellenberg, Jordan. 2014. *How Not to Be Wrong: The Power of Mathematical Thinking.* New York: Hudson.

Ernest, Paul. 1998. "The Culture of the Mathematics Classroom and the Relations Between Personal and Public Knowledge: An Epistemological Perspective." In *The Culture of the Mathematics Classroom,* edited by Falk Seeger, Jörg Voigt, and Ute Waschescio, 245–68. Cambridge: Cambridge University Press.

———. 1999a. "Forms of Knowledge in Mathematics and Mathematics Education: Philosophical and Rhetorical Perspectives." *Educational Studies in Mathematics* 38:67–83.

———. 1999b. "Is Mathematics Discovered or Invented?" *Philosophy of Mathematics Education* 12:n.p.

Gardner, David P. 1983. *A Nation at Risk: The Imperative for Educational Reform.* Washington, DC: National Commission on Excellence in Education.

Garland, Sarah. 2014a. "Who Was Behind the Common Core Math Standards, and Will They Survive?" Last modified December 29. http://hechingerreport.org/behind-common-core-math-standards-will-survive.

———. 2014b. "Why Is This Common Core Math Problem So Hard? Supporters Respond to Quiz That Went Viral." Last modified March 26. http://hechingerreport.org/common-core-math-problem-hard-supporters-common-core-respond-problematic-math-quiz-went-viral.

Germaine-McCarthy, Yvelyne. 1999. *Bringing the* NCTM *Standards to Life: Exemplary Practices from High Schools.* New York: Routledge.

Goodnight, G. Thomas. 1982. "The Personal, Technical, and Public Spheres of Argumentation." *Journal of the American Forensics Association* 18:214–27.

Green, Elizabeth. 2014. "Why Do Americans Stink at Math?" *New York Times Magazine,* July 23. https://www.nytimes.com/2014/07/27/magazine/why-do-americans-stink-at-math.html.

Haas, Angela M. 2012. "Race, Rhetoric, and Technology: A Case Study of Decolonial Technical Communication Theory, Methodology, and Pedagogy." *Journal of Business and Technical Communication* 26:277–310.

Hayden, Robert W. 1981. "A History of the 'New Math' Movement in the United States." Retrospective Theses and Dissertations, Iowa State University, Paper 7427.

Hekimoglu, Serkan, and Margaret Sloan. 2005. "A Compendium of Views on the NCTM Standards." *Mathematics Educator* 15:35–43.

Hellman, Geoffrey, and John L. Bell. 2006. "Pluralism and the Foundations of Mathematics." In *Scientific Pluralism,* edited by Stephen H. Kellert, Helen E. Longino, and C. Kenneth Waters, 64–79. Minnesota Studies in the Philosophy of Science 19. Minneapolis: University of Minnesota Press.

Herbel-Eisenmann, Beth, Nathalie Sinclair, Kathryn B. Chval, Marta Civil, Stephen J. Pape, Michelle Stephan, Jeffrey J. Wanko, and Trena L. Wilkerson. 2016. "Positioning Mathematics Education Researchers to Influence Storylines." *Journal for Research in Mathematics Education* 47:102–17.

Herrera, Terese A., and Douglas T. Owens. 2001. "The 'New New Math'? Two Reform Movements in Mathematics Education." *Theory into Practice* 40 (2): 84–92.

Holmes, Patrick. 2012. "*A Nation at Risk* and Education Reform: A Frame Analysis." MA thesis, University of Washington. https://digital.lib.washington.edu/researchworks/bitstream/handle/1773/21944/Holmes_washington_0250O_10829.pdf.

Izmirli, Ilhan M. 2011. "Pedagogy on the Ethnomathematics-Epistemology Nexus: A Manifesto." *Journal of Humanistic Mathematics* 1:27–50.

Katz, Brian. 2017. "Supremum/Supremacy." Inclusion/Exclusion (blog). American Mathematical Society, May 4. http://blogs.ams.org/inclusionexclusion/2017/05/04/supremumsupremacy.

Kenney, Joan M. 2005. "Mathematics as Language." In *Literacy Strategies for Improving Mathematics Instruction,* edited by Joan M. Kenney, Euthecia Hancewicz, Loretta Heuer, Diana Metsisto, and Cynthia L. Tuttle, 1–8. Alexandria, VA: Association for Supervision and Curriculum Development.

Klein, David. 2003. "A Brief History of American K-12 Mathematics Education in the 20th Century." In *Mathematical Cognition,* edited by James M. Royer, 175–226. Greenwich, CT: Information Age.

Krips, Henry, J. E. McGuire, and Trevor Melia. 1995. Introduction to *Science, Reason, and Rhetoric,* edited by Henry Krips, J. E. McGuire and Trevor Melia, vii–xix. Pittsburgh: University of Pittsburgh Press.

Lahman, Mary P., and Alison Lietzenmayer. 2015. "Work-Life and the Popular Press: How Words Create Worlds." *ETC: A Review of General Semantics* 72:127–48.

Lerman, Stephen. 1998. "Cultural Perspectives on Mathematics and Mathematics Teaching and Learning." In *The Culture of the Mathematics Classroom*, edited by Falk Seeger, Jörg Voigt, and Ute Waschescio, 290–307. Cambridge: Cambridge University Press.

Lyne, John, and Henry F. Howe. 1986. "'Punctuated Equilibria': Rhetorical Dynamics of a Scientific Controversy." *Quarterly Journal of Speech* 72:132–47.

Maddy, Penelope. 1989. "The Roots of Contemporary Platonism." *Journal of Symbolic Logic* 54 (4): 1121–44.

Martin, Danny Bernard. 2008. "E(race)ing Race from a National Conversation on Mathematics Teaching and Learning: The National Mathematics Advisory Panel as White Institutional Space." *Mathematics Enthusiast* 5 (2): 387–97. https://scholarworks.umt .edu/tme/vol5/iss2/20.

———. 2013. "Race, Racial Projects, and Mathematics Education." *Journal for Research in Mathematics Education* 44 (1): 316–33.

———. 2015. "The Collective Black and *Principles to Actions*." *Journal of Urban Mathematics Education* 8 (1): 17–23.

Martin, Danny Bernard, Maisie L. Gholson, and Jacqueline Leonard. 2011. "Mathematics as Gatekeeper: Power and Privilege in the Production of Knowledge." *Journal of Urban Mathematics Education* 3 (2): 12–24.

NCTM (National Council of Teachers of Mathematics). 1980. *An Agenda for Action: Directions for School Mathematics for the 1980s*. Reston, VA: National Council of Teachers of Mathematics.

———. 1989. *Curriculum and Evaluation Standards for School Mathematics*. Reston, VA: National Council of Teachers of Mathematics.

———. 2000. *Principles and Standards for School Mathematics*. Reston, VA: National Council of Teachers of Mathematics.

———. 2006. *Curriculum Focal Points for Pre-K–Grade 8 Mathematics*. Reston, VA: National Council of Teachers of Mathematics.

———. 2014. *Principles to Actions: Ensuring Mathematical Success for All*. Reston, VA: NCTM.

NCTM et al. (National Council of Teachers of Mathematics, Mathematics Center for Science, and Engineering Education, and National Research Council). 1997. *Improving Student Learning in Mathematics and Science: The Role of National Standards in State Policy*. Washington, DC: National Academy Press.

Newsweek. 2007. "Math Makeover: It Adds Up for Girls." Last modified August 5. https://www.newsweek.com/math-makeover-it-adds-girls-99533.

Noddings, Nel. 1990. "Constructivism in Mathematics Education." *Journal for Research in Mathematics Education* 4:7–18.

Park, Jennifer. 2004. "A Nation at Risk." *Education Week*, September 10. http://www.edweek .org/ew/issues/a-nation-at-risk.

Paulos, John Allen. 1988. *Innumeracy: Mathematical Illiteracy and Its Consequences*. New York: Hill and Wang.

Podnieks, Karlis. 2015. "Fourteen Arguments in Favour of a Formalist Philosophy of Real Mathematics." *Baltic Journal of Modern Computing* 3:1–15.

Pollak, Henry. 1983. *The Mathematical Sciences Curriculum K–12: What Is Still Fundamental and What Is Not*. Washington, DC: Conference Board of the Mathematical Sciences.

Pugalee, David K. 1999. "Constructing a Model of Mathematical Literacy." *Clearing House* 73:19–22.

Rappaport, David. 1971. "The Nuffield Mathematics Project." *Elementary School Journal* 71 (6): 295–308.

Reyes, G. Mitchell. 2004. "The Rhetoric in Mathematics: Newton, Leibniz, the Calculus, and the Rhetorical Force of the Infinitesimal." *Quarterly Journal of Speech* 90:163–88.

———. 2014. "Stranger Relations: The Case for Rebuilding Commonplaces Between Rhetoric and Mathematics." *Rhetoric Society Quarterly* 44:470–91.

Roberts, Rebecca Anne. 2015. "Parents and the Common Core State Standards for Mathematics." MA thesis, Brigham Young University. https://scholarsarchive.byu.edu/etd/4396.

Romberg, Thomas. 1993. "NCTM's Standards: A Rallying Flag for Mathematics Teachers." *Educational Leadership* 50 (5): 36–41.

———. 2017. "Changes in School Mathematics: Curricular Changes, Instructional Changes, and Indicators of Change." *CPRE Research Reports*, August 21. https://repository.upenn.edu/cpre_researchreports/105.

Ross, Kenneth A. 2000. "The MAA and the New NCTM Standards." Mathematical Association of America. https://www.maa.org/the-maa-and-the-new-nctm-standards.

Rotman, Brian. 1993. *Ad Infinitum . . . : the Ghost in Turing's Machine: taking God Out of Mathematics and Putting the Body Back In: an Essay in Corporeal Semiotics*. Stanford: Stanford University Press.

———. 2000. *Mathematics as Sign: Writing, Imagining, Counting (Writing Science)*. Stanford: Stanford University Press.

Schoenfeld, Alan H. 2004. "The Math Wars." *Educational Policy* 18:253–86.

Seeger, Falk. 1998. "Representations in the Mathematics Classroom: Reflections and Constructions." In *The Culture of the Mathematics Classroom*, edited by Falk Seeger, Jörg Voigt, and Ute Waschescio, 308–43. Cambridge: Cambridge University Press.

Simon, Martin A. 1995. "Reconstructing Mathematics Pedagogy from a Constructivist Perspective." *Journal for Research in Mathematics Education* 26:114–45.

Solomon, Yvette. 2009. *Mathematical Literacy*. New York: Routledge.

Steinbring, Heinz. 1998. "Mathematical Understanding in Classroom Interaction: The Interrelation of Social and Epistemological Constraints." In *The Culture of the Mathematics Classroom*, edited by Falk Seeger, Jörg Voigt, and Ute Waschescio, 344–72. Cambridge: Cambridge University Press.

Steudeman, Michael J. 2014. "'The Guardian Genius of Democracy': The Myth of the Heroic Teacher in Lyndon B. Johnson's Education Policy Rhetoric, 1964–1966." *Rhetoric and Public Affairs* 17:477–510.

Tobias, Sheila. 1978. *Overcoming Math Anxiety*. Boston: Houghton Mifflin.

Tobias, Sheila, and Carol Weissbrod. 1980. "Anxiety and Mathematics: An Update." *Harvard Educational Review* 50:63–70.

Van Amerom, Barbara A. 2003. "Focusing on Informal Strategies when Linking Arithmetic to Early Algebra." *Educational Studies in Mathematics* 54:63–75.

Van der Blij, Frederik, Sven Hilding, and Ari I. Weinzweig. 1980. "A Synthesis of National Reports on Changes in Curricula." In *Comparative Studies of Mathematics Curricula: Change and Stability, 1960–1980*, edited by Hans Steiner, 37–54. Bielefeld, Germany: Institut für Didaktik der Mathematik der Universität Bielefeld.

Wilder, Raymond L. 1981. "Examples of Cultural Patterns Observable in the Evolution of Mathematics." In *Mathematics as a Cultural System*, 21–46. New York: Pergamon.

Woodward, John. 2004. "Mathematics Education in the United States: Past to Present." *Journal of Learning Disabilities* 37:16–31.

Contributors

Catherine Chaput is Professor of English at the University of Nevada, Reno. Her research focuses on rhetoric, political economy, and affect studies. She has published articles and book chapters that address the diverse rhetorical strategies of economic policy. Her first book, *Inside the Teaching Machine* (2008), tracks the political, economic, and culture circulation of higher education discourse as foundational to the globalization of market-driven university models. Her most recent book, *Market Affect and the Rhetoric of Political Economic Debates* (2019), explores capitalism and its critics at key historical moments. It ends by suggesting the invention of new affective modalities as crucial to anticapitalist politics.

Crystal Broch Colombini is Assistant Professor of English at Fordham University. Her work asks how public rhetoric and writing shapes and responds to economic circumstances, particularly cultural, social, and political transformations around the rise and reign of neoliberal capitalism. Her first book project, *Constructing the Market: Rhetoric, Risk, and American Homeownership* (in process), seeks to understand how historical and contemporary rhetoric about homeownership authorizes political economic transformation in the United States. Her articles have appeared in *Rhetoric Society Quarterly*, *Advances in the History of Rhetoric*, and *College English*.

Nathan Crick is a Professor of Communication at Texas A&M. His work explores the relationship between rhetoric and power in different periods of political and social change. His books *Democracy and Rhetoric: John Dewey on the Arts of Becoming* (2010) and *Dewey for a New Age of Fascism: Teaching Democratic Habits* (2019) constructs a view of rhetoric, logic, and aesthetics that is consistent with an ethics of democracy that promotes creative individuality. His books *Rhetoric and Power: The Drama of Classical Greece* (2014) and *The Keys of Power: The Rhetoric and Politics of Transcendentalism* (2017) explore how major historical figures conceptualize the function of rhetoric in history.

Michael Dreher is a Professor of Communication Studies at Bethel University. His teaching and research explore the rhetoric of inquiry, including mathematics, science, and religion. He is particularly interested in the ways in which the K-12 education system influences how students and teachers learn and understand how to communicate in various disciplines.

Jeanne Fahnestock is Professor Emeritus at the University of Maryland. Her work applies the rhetorical tradition to science, emphasizing the links between argument, style, and visualization. In addition to articles and chapters, her first book, *A Rhetoric of Argument*, written with Marie Secor (2004, 3rd ed.), revises stasis theory for teaching argumentation. Her second book, *Rhetorical Figures in Science* (1999, 2002), examines arguments in science epitomized by neglected figures of speech. Her third book, *Rhetorical Style* (2011), combines rhetorical stylistics with linguistic approaches to promote the study of persuasive language.

Andrew C. Jones is an Assistant Professor of Communications at LCC International University, Latvia. His research interests include rhetoric and detective fiction, political communication, and presidential address in Eastern Europe.

Joseph Little is an Associate Professor of English at Niagara University, where he directs the university's first-year writing program. His recent publications include *Letters from the Other Side of Silence* (2017) and "Analogy in William Rowan Hamilton's New Algebra" (*Technical Communication Quarterly* 2012), which he cowrote with mathematician Maritza Branker. His current work explores the role of imagery in the reception of Abraham Maslow's hierarchy of needs.

G. Mitchell Reyes is an Associate Professor at Lewis and Clark College. His work examines the interfaces, historical and contemporary, between rhetoric and mathematics. From his early work on the role of rhetoric in the invention of the calculus ("The Rhetoric in Mathematics: Newton, Leibniz, the Calculus, and the Rhetorical Force of the Infinitesimal," *Quarterly Journal of Speech* 2004) to his more recent work on the rhetorical force of algorithms and algorithmic culture ("Algorithms and Rhetorical Inquiry: The Case of the 2008 Financial Collapse," *Rhetoric and Public Affairs* 2019), Reyes consistently seeks to understand how mathematical discourse shapes both humans and nonhumans into increasingly complex and novel configurations.

Edward Schiappa is John E. Burchard Professor of Humanities at the Massachusetts Institute of Technology. His work on classical and contemporary rhetorical and argumentation theory has appeared in many scholarly journals, and he is author of *Defining Reality: Definitions and the Politics of Meaning* (2003), *Protagoras and Logos: A Study in Greek Philosophy and Rhetoric* (1991), *The Beginnings of Greek Rhetorical Theory* (1999), *Beyond Representational Correctness: Rethinking Criticism of Popular Media* (2008), coauthor (with David M. Timmerman) of *Classical Greek Rhetorical Theory and the Disciplining of Discourse* (2010) and (with John P. Nordin) *Argumentation: Keeping Faith with Reason* (2014), editor of *Warranting Assent: Case Studies in Argument Evaluation* (1995), and former editor of *Argumentation and Advocacy*.

James Wynn is an Associate Professor of English and Rhetoric at Carnegie Mellon University. His work examines the intersections of the rhetoric, science, and mathematics. His first book, *Evolution by the Numbers: The Origin of Mathematics in Biology* (2012), investigates how mathematical warrants become accepted sources for argument in the biological sciences. His second, *Citizen Science in the Digital Age* (2017), examines how the technical affordances of the internet and internet-connectable devices are changing the relationships between science, scientists, nonexperts, and policy-makers. He has also published articles in *Written Communication*, *Rhetorica*, and *Nineteenth-Century Prose*.

Index

Italicized page references indicate illustrations. Endnotes are referenced with "n" followed by the endnote number.